I0473876

The Complete

Calculus Review Book

Henry Gu

Author: Henry Gu
Editor: Christopher Gu

ISBN-10: 1478362200   ISBN-13: 978-1478362203

"Calculus is the subject.
Function is the object.
Infinity is the concept.
Limit is the result."

Preface

This book is for math teachers and professors who need a handy calculus reference book, for college students who need to master the essential calculus concepts and skills, and for AP Calculus students who want to pass the exam with a perfect score.

Calculus can not be made easy, but it can be made simple.
This book is concise, but the scope of the contents is not.

To solve calculus problems, you need strong algebra skills. The only way to build these skills is through practice. To practice, you need this book.

Acknowledgement

Thanks to my family for all their love and support.

Thanks to you, the reader, for choosing this book as your study guide. I wish you the best of luck!

Contents

I. Functions   1.
  1. Relation and Function   1.
    Vertical Line Test, Horizontal Line Test, One-to-One Function   1.
    Domain and Range, Interval Notation, Function Notation   2.
  2. Composition of Functions, Decomposition of Functions   2.
  3. The Domain of the Composition of Functions, Inverse Functions   3.
  4. Function under Transformations   4.
    Translation, Reflection, Dilation   4.
  5. Increasing, Decreasing, and Constant Functions   5.
  6. Odd and Even Functions   5.
  7. Periodic Functions   6.
  8. Extreme Values of Function, The Extreme Value Theorem   6.
    Local Maximum, Local Minimum, Absolute Maximum, Absolute Minimum   6.
  9. Asymptotes of Functions   7.

II. Function Graphs   8.
  1. Rectangular Coordinates   8.
    (1) Slope, Midpoint, and Distance (2) Linear Function   8.
    (3) Quadratic Functions and Parabolas (4) Absolute Value Functions   8.
    (5) Square Root Function (6) Inverse Variation   9.
    (7) Exponential Function (8) Logarithmic Function (9) Greatest Integer Function   10.
    (10) Equation of a Circle (11) Ellipses, Circles (12) Hyperbolas   11.
    (13) Trigonometric Functions and Their Inverse Functions   12.
  2. Parametric Equations, The Inverse Function in Parametric Forms   15.
  3. Polar Coordinates   16.
    (1) Line (2) Circles   16.
    (3) Limaçon Curves: Cardioid Curves, Dimpled Limaçon,   17.
      Limaçon with an Inner Loop (4) Rose Curves   18.
    Symmetry of Polar Graphs, Converting to and from Rectangular   19.

III. Limits and Continuity   20.
  1. Limits   20.
  2. Continuity of Functions and Composition of Functions   20.
  3. Infinitesimal and Infinity   22.
    The Order of Infinitesimals and Infinities   22.
  4. Finding Limits I.   23.
    (1) Plug in (2) Use Equivalent Infinitesimals   23.
    (3) Use Equivalent Infinities (4) Use Substitution (5) Eliminate Subtraction   24.
    (6) Composition of Functions (7) Some Important Limits   25.

IV. Differentiation   26.
  1. Derivative   26.
  2. Differentiability   27.
  3. Derivative of Composite Functions   28.

4. Implicit Differentiation   28.
5. Derivative of Inverse Functions   28.
6. Derivative of Parametric Functions   30.
7. Derivative of Polar Functions   31.

V. Applications of Derivatives   **32.**
1. Curve Sketching   32.
2. Maximum and Minimum Problems   33.
3. Related Rates   34.
4. Average Rate of Change   35.
5. Linear Approximation   35.
6. Finding Limits II.   36.
   (1) Use Derivative Definition (2) Indeterminate Forms of Limits:
   $\dfrac{0}{0}$, $\dfrac{\infty}{\infty}$, $0 \bullet \infty$, $\infty - \infty$, 0^0 , 1^∞ , ∞^0

VI. Antiderivatives   **38.**
1. Antiderivative   38.
2. Rules of Integation   38.
3. Substitution Rule, Trigonometric Substitution   38.
4. Using Differential Identities   39.
5. Using Trigonometric Identities   40.
6. Completing the Square   41.
7. Integration by Parts   41.
8. Integration by Partial Fractions   42.

VII. Definite Integrals   **44.**
1. Definite Integral   44.
2. Calculating the Definite Integral   44.
   (1) Riemann Sum   44.
   Riemann Sum for Approximation , Trapezoid Rule for Approximation   45.
   (2) The Fundamental Theorem of Calculus, Part 1   46.
3. Rules of Definite Integration   46.
4. Substitution Rule   46.
5. Definite Integrals by Parts   47.
6. Definite Integrals of Parametric Functions   47.
7. The Fundamental Theorem of Calculus, Part 2   48.
8. Improper Definite Integrals   49.
   Type1: The interval of the integration is infinite   49.
           Comparison Test, Limit Comparison Test   50.
   Type2: The function f(x) has infinite discontinuities in the interval   52.

VIII. Applications of Integration   **54.**
  1. Area   54.
  2. Volume   57.
     (1) Solids with Known Cross Section Area   57.
     (2) Solids of Revolution   58.
  3. Arc Length   60.
  4. Surface Area of Revolution   63.
  5. Work   64.
  6. Variable-Density Mass   65.
  7. The Net Change   66.

IX. Motion and Vectors   **67.**
  1. Motion in One-Dimension   67.
  2. Motion in Two-Dimension   70.
     (1) Vectors   70.
     (2) Position Vector,Velocity Vectors and Acceleration Vectors   74.

X. Mean Value Theorems   **78.**
  1. The Intermediate Value Theorem   78.
  2. The Mean Value Theorem, Rolle's Theorem   78.
  3. Average Value (The Mean Value Theorem for Integrals)   80.

XI. Differential Equations   **81.**
  1. Separation of Variables   81.
  2. Homogeneous Equations   82.
  3. Linear First-Order Equations   83.
  4. Exponential Growth and Decay, Newton's Law of Cooling, The Logistic Model   84.
  5. Euler's Method (Linear Approximation)   88.
  6. Slope Field   88.

XII. Infinite Series   **90.**
  Part I. Series with Constant Terms   90.
  1. The nth Term Test   91.
  2. Geometric Series Test   91.
  3. P-Series Test   92.
  4. Integral Test   93.
  5. Comparison Test, The Limit Comparison Test   94.
  6. Alternating Series Test   95.
  7. The Absolute Convergence Test   95.
  8. Ratio Test   95.
  9. The nth-Root Test   96.

Part II. Series with Variable Terms .... 97.
1. Power Series, General Form of the Power Series .... 97.
   Term-by-Term Differentiation and Integration of the Power Series ... 99.
2. Functions Represented as a Power Series .... 101.
   Taylor Series, Maclaurin Series, ... 101.
   Lagrange Error Bound, ... 102.
   Alternating Series Estimation Theorem .... 103.
   Some Common Functions Defined by Power Series .... 103.

Appendices

A1. Basic Formulas .... **104.**
1. Algebraic Formulas .... 104.
2. Geometry Formulas .... 105.
3. Trigonometry Formulas .... 106.

A2. Calculus Formulas .... **108.**
1. Rules of Differentiation, Rules of Integation, Rules of Definite Integration ... 108.
2. Basic Derivative Table .... 109.
3. Basic Integral Table .... 110.
4. Infinite Series .... 111.
   Some Common Functions Defined by Power Series .... 111.

A3. Using Graphing Calculator .... **112.**
Part 1. Basic Operations .... 112.
1. Clear the Memory, Return to Home Screen .... 112.
2. Graph Functions .... 112.
   Zoom Menu, Change Window Dimensions, Trace .... 112.
3. Table of a Function .... 112.
4. Calculations .... 112.
   Solve Equations, Solve the System of Equations, Maxmum and Minimum ... 112.
Part 2. Calculus Applications .... 113.
1. Graph Piecewise Defined Functions .... 113.
2. Find the Numerical Derivatives .... 113.
3. Find the Numerical Integral .... 113.
4. Sequence and Series .... 113.

1. Relation and Function

In the coordinate plane, every point corresponds to one pair of ordered numbers (x, y), and vice versa.

A **relation** is a set of points, which can be continuous or scattering.

A **function** is a special relation, which passes the vertical line test.

Vertical Line Test:
If any vertical line intersects the graph at only one point, then the relation is a function.

   e.g. $y^2 = x$ is equivalent to $y = \pm\sqrt{x}$. It is not a function.

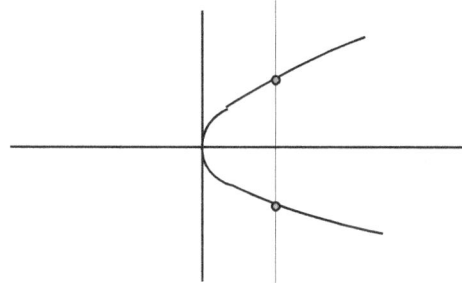

A **one-to-one function** is a special function, which passes both vertical line test and horizontal line test.

Horizontal Line Test:
If any horizontal line intersects the graph at only one point, then the function is a **one-to-one function**.

   e.g. $y = x^2$ is a function, but not a one-to-one function.

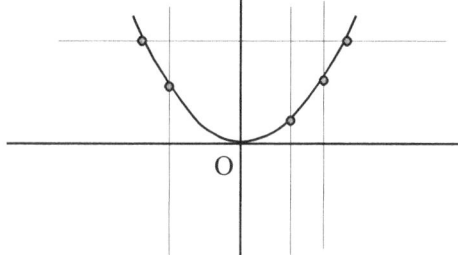

   If we restrict the domain of $y = x^2$ to $x \geq 0$,
   then the function is a one-to-one funcion.

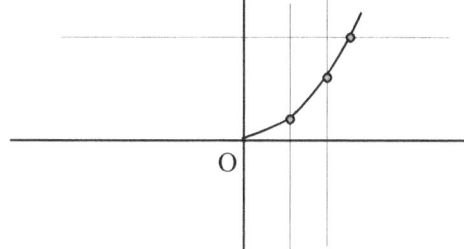

2.

I. Functions

Domain and Range

e.g. $y = x^2$ Domain: $\{x \mid x \text{ all real numbers}\}$,   Range: $\{y \mid y \geq 0\}$

e.g. $y = \sqrt{x}$ Domain: $\{x \mid x \geq 0\}$,   Range: $\{y \mid y \geq 0\}$

e.g. $y = \dfrac{1}{x^2 - 9}$ Domain: $\{x \mid x \text{ all real numbers except } \pm 3\}$

e.g. $y = \dfrac{1}{\sqrt{x - 3}}$ Domain: $\{x \mid x > 3\}$

Interval Notation

e.g. $(2, 5)$ represents $\{ x \mid 2 < x < 5 \}$

$[2, 5]$ represents $\{ x \mid 2 \leq x \leq 5 \}$

$(2, 5]$ represents $\{ x \mid 2 < x \leq 5 \}$

$[2, 5)$ represents $\{ x \mid 2 \leq x < 5 \}$

e.g. $(-\infty, \infty)$ represents $\{ x \mid x \text{ all real numbers} \}$

$(-\infty, -5)$ represents $\{ x \mid x < -5 \}$

$[5, \infty)$ represents $\{ x \mid x \geq 5 \}$

$(-\infty, -5) \cup [5, \infty)$ represents $\{ x < -5 \text{ or } x \geq 5 \}$

Function Notation

e.g. $y = \dfrac{2x}{x - 1}$, its function notation: $f(x) = \dfrac{2x}{x - 1}$ or $y(x) = \dfrac{2x}{x - 1}$

$f(5) = \dfrac{2 \cdot 5}{5 - 1} = \dfrac{10}{4} = \dfrac{5}{2}$

$f(a + 2) = \dfrac{2(a + 2)}{(a + 2) - 1} = \dfrac{2a + 4}{a + 1}$

$f(x^2) = \dfrac{2(x^2)}{(x^2) - 1} = \dfrac{2x^2}{x^2 - 1}$

2. Composition of Functions

$(f \circ g)(x) = f(g(x))$

e.g. $f(x) = x^2 - 1$, $g(x) = x + 1$

$(f \circ g)(x) = f(g(x)) = f(x + 1) = (x + 1)^2 - 1 = x^2 + 2x$

$(f \circ g)(2) = f(g(2)) = f(2 + 1) = f(3) = 3^2 - 1 = 8$

$(g \circ f)(x) = g(f(x)) = g(x^2 - 1) = (x^2 - 1) + 1 = x^2$

$(g \circ f)(2) = g(f(2)) = g(2^2 - 1) = g(3) = 3 + 1 = 4$

Note: $(f \circ g)(x) \neq (g \circ f)(x)$

Decomposition of Functions

e.g. $y = 3\sin^2 x$

$y = f(u) = 3u^2$, $u(x) = \sin x$

e.g. $y = \ln(4 - x^2)$

$y = f(u) = \ln u$, $u(x) = 4 - x^2$

The Domain of the Composition of Functions

$$(f \circ g)(x) = f(g(x))$$

The domain of f restricts the range of g, which restricts the domain of g. Therefore the domain of $f \circ g$ is the set of all x in the domain of g such that the range of g(x) is in the domain of f.

e.g. $f(x) = \sqrt{x}$ and $g(x) = \ln x$
    $(f \circ g)(x) = f(g(x)) = \sqrt{\ln x}$

The domain of f restricts the range of g such as $\ln x \geq 0$, therefore $x \geq 1$ is the domain of $f \circ g$.

3. Inverse Functions

For every one-to-one function f(x), there is an inverse function $f^{-1}(x)$.
For function y = f(x), its inverse is x = f(y) or denoted as $y = f^{-1}(x)$.

e.g. $y = x^2$
It is a function.   (passing the vertical line test)
But it is not a one-to-one function. (failing the horizonal line test)

e.g. $y = f(x) = e^x$, its inverse is $x = e^y$ (implicit form)
                  or $y = \ln x$ (explicit form).

The domain of the inverse function is the range of the original function.

e.g.   Original $f(x) = \{(1, 1), (2, 4), (3, 9)\}$ Domain: $\{x \mid x = 1, 2, 3\}$, Range: $\{y \mid y = 1, 4, 9\}$
    Inverse $f^{-1}(x) = \{(1, 1), (4, 2), (9, 3)\}$ Domain: $\{x \mid x = 1, 4, 9\}$, Range: $\{y \mid y = 1, 2, 3\}$

The composition of a function and its inverse:
$$f(f^{-1}(x)) = x$$

e.g. $\ln e^x = x$, $e^{\ln x} = x$

Finding the Inverse Function

e.g. $f(x) = 3x + 5$ find $f^{-1}(x)$
    $y = 3x + 5$ express f(x) as y
    $x = 3y + 5$ interchange x and y
    $y = \dfrac{x - 5}{3}$ solve for y in terms of x

    $f^{-1}(x) = \dfrac{x - 5}{3}$

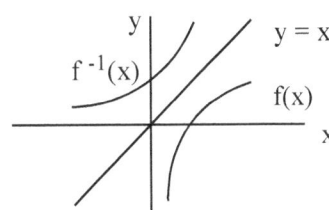

The graph of $f^{-1}(x)$ is the reflection of f(x) in the y = x.

4. Function under Transformations

Translation: $y = f(x) \xrightarrow{\ \ T_{a,b}\ \ } y = f(x-a) + b$

Reflection: $y = f(x) \xrightarrow{\ \ r_{x-axis}\ \ } y = -f(x)$

$y = f(x) \xrightarrow{\ \ r_{y-axis}\ \ } y = f(-x)$

Dilation: $y = f(x) \xrightarrow[\text{vertical shrink if } 0<a<1]{\text{vertical stretch if } a>1} y = af(x)$

Tip: Use the graphing calculator to verify the answer.

e.g.

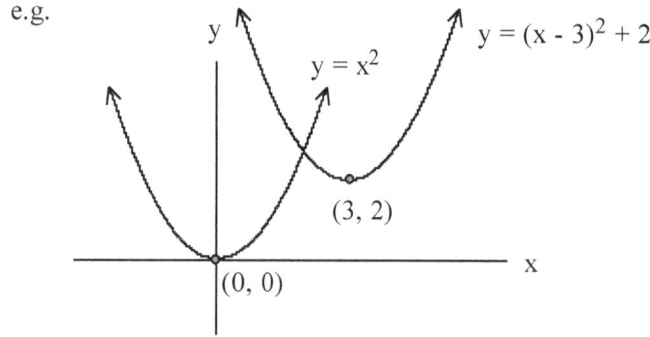

$y = x^2 \xrightarrow{\ \ T_{3,2}\ \ } y = (x-3)^2 + 2$

e.g.

$y = \sqrt{x} \xrightarrow{\text{moved 5 units to the left}} y = \sqrt{x+5}$

$y = \sqrt{x+5} \xrightarrow{\ \ r_{x-axis}\ \ } y = -\sqrt{x+5}$

e.g.

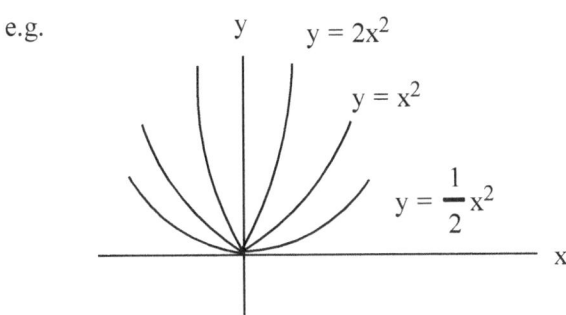

$$y = x^2 \xrightarrow{\text{vertical stretch by a factor of 2}} y = 2x^2$$

$$y = x^2 \xrightarrow{\text{vertical shrink by a factor of } \frac{1}{2}} y = \frac{1}{2}x^2$$

5. Increasing, Decreasing, and Constant Functions

When $a < b$

$f(a) < f(b)$ Increasing function

$f(a) > f(b)$ Decreasing function

$f(a) = f(b)$ Constant function

e.g. $y = x^2$, Increasing on $(0, \infty)$ and decreasing on $(-\infty, 0)$.

6. Odd and Even Functions

An odd function is symmetric about the origin.
$$f(-x) = -f(x)$$

e.g. $f(x) = 2x^3$, $f(-x) = 2(-x)^3 = -2x^3 = -f(x)$

An even function is symmetric about the y-axis.
$$f(-x) = f(x)$$

e.g. $f(x) = 3x^2 + 5$, $f(-x) = 3(-x)^2 + 5 = 3x^2 + 5 = f(x)$

A constant is an even function.
The sum of odd functions is an odd function.
The sum of even functions is an even function.
The product of two odd functions is an even function.
The product of two even functions is an even function.
The product of an odd function and an even function is an odd function.

I. Functions

7. Periodic Functions

A function f(x) is periodic if there is a positive number p such that $f(x + p) = f(x)$
for every value of x in the domain. The smallest such number p is the period of the function.

y = asinbx   and   y = acosbx

amplitude $= |a|$, frequency $= |b|$, period $= \dfrac{2\pi}{|b|}$

e.g. $y = -3\sin2x$,   amplitude $= 3$, frequency $= 2$, period $= \dfrac{2\pi}{2} = \pi$

$y = 5\sin3(x + \dfrac{\pi}{4}) + 1$,   amplitude $= 5$, frequency $= 3$, period $= \dfrac{2\pi}{3}$

y = tanbx

frequency $= |b|$, period $= \dfrac{\pi}{|b|}$

e.g. $y = \tan x$ period $= \pi$, $y = \tan2x$ period $= \dfrac{\pi}{2}$

8. Extreme Values of Functions

A value f(c) is a **local maximum (relative maximum)** of f(x) if there is an open interval (a, b)
containing a value c such that $f(x) \le f(c)$ for all values of x in (a, b).

A value f(d) is a **local minimum (relative minimum)** of f(x) if there is an open interval (a, b)
containing a value d such that $f(x) \ge f(d)$ for all values of x in (a, b).

The **absolute maximum** is the largest value of f(x) in its domain.
The **absolute minimum** is the smallest value of f(x) in its domain.

The Extreme Value Theorem

If f is continuous on the closed interval [a, b], then there exist an absolute maximum value M and an
absolute minimum value of m somewhere in the interval. That is, for all x in the domain

$$m \le f(x) \le M$$

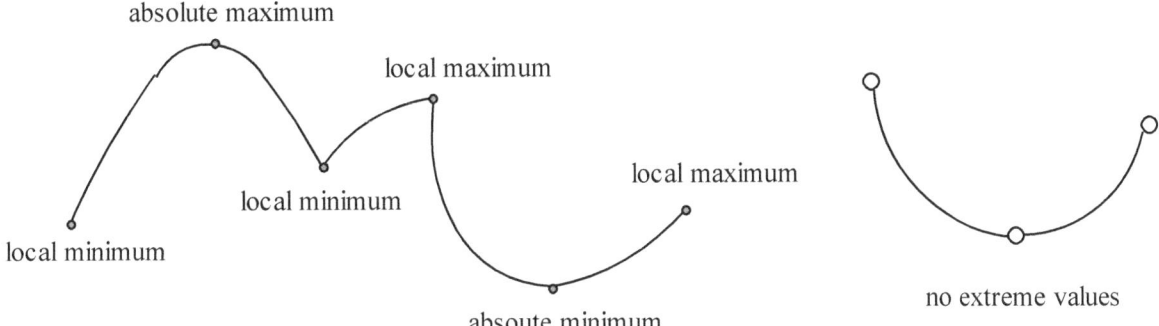

*9. Asymptotes of Functions

The horizontal line y = b is a **horizontal asymptote** of a function y = f(x), if

$$\lim_{x \to \infty} f(x) = b \qquad \text{or} \qquad \lim_{x \to -\infty} f(x) = b$$

The vertical line x = a is a **vertical asymptote** of a function y = f(x), if

$$\lim_{x \to a^+} f(x) = \pm \infty \qquad \text{or} \qquad \lim_{x \to a^-} f(x) = \pm \infty$$

The line y = mx + b is an **oblique asymptote (slant asymptote)** of a function y = f(x), if

$$\lim_{x \to \infty} f(x) = mx + b \qquad \text{or} \qquad \lim_{x \to -\infty} f(x) = mx + b$$

e.g. Find asymptotes of function $f(x) = \dfrac{x+2}{x+5}$

$$\text{Rewrite } f(x) = \frac{x+5-3}{x+5} = 1 - \frac{3}{x+5} \qquad \text{quotient and remainder}$$

$$\lim_{x \to \infty} f(x) = 1$$

$$\lim_{x \to -5} f(x) = \pm \infty$$

Therefore y = 1 is a horizontal asymptote and x = −5 is a vertical asymptote.

e.g. Find asymptotes of function $f(x) = \dfrac{x^2 - x - 1}{x - 2}$

$$f(x) = \frac{x^2 - x - 1}{x - 2} = x + 1 + \frac{1}{x - 2} \qquad \text{quotient and remainder}$$

$$\lim_{x \to \pm \infty} \frac{1}{x - 2} = 0 \qquad \text{y = x + 1 is an oblique asymptote.}$$

$$\lim_{x \to 2} f(x) = \pm \infty \qquad \text{x = 2 is a vertical asymptote.}$$

* Read this topic after Chapter III. Limits.

II. Function Graphs

1. Rectangular Coordinates

(1) Slope, Midpoint, and Distance

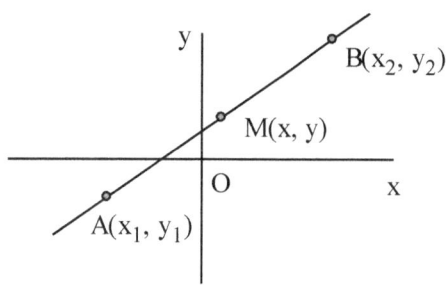

slope $\quad m = \dfrac{y_2 - y_1}{x_2 - x_1}$

Two lines are parallel: $m_1 = m_2$

Two lines are perpendicular (normal): $m_2 = -\dfrac{1}{m_1}$

midpoint $\quad M(\bar{x}, \bar{y}) = M(\dfrac{x_1 + x_2}{2}, \dfrac{y_1 + y_2}{2})$

distance $\quad d = \sqrt{(x_2 - x_1)^2 + (y_2 - y_1)^2}$

(2) Linear Function

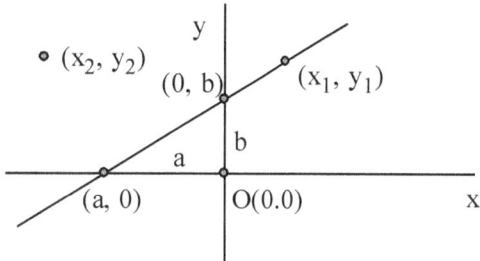

Slope-Intercept Form: $\quad y = mx + b$

Point-Slope Form: $\quad y - y_1 = m(x - x_1)$

Intercept Form: $\quad \dfrac{x}{a} + \dfrac{y}{b} = 1$

General Form: $\quad Ax + By + C = 0$

Distance from point (x_2, y_2) to line $Ax + By + C = 0$

$$d = \left| \dfrac{Ax_2 + By_2 + C}{\sqrt{A^2 + B^2}} \right|$$

(3) Quadratic Functions and Parabolas

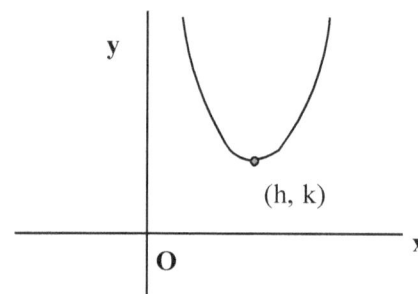

General Form: $y = f(x) = ax^2 + bx + c \quad$ where $a \neq 0$

Vertex Form: $\quad y = f(x) = a(x - h)^2 + k$

Axis of Symmetry: $x = h = -\dfrac{b}{2a}$

Vertex (Turning Point): (h, k)

$$h = -\dfrac{b}{2a} \quad , \quad k = f\left(-\dfrac{b}{2a}\right)$$

(4) Absolute Value Functions

$y = |x|$

when $x < 0$
$y = -x$

when $x \geq 0$,
$y = x$

e.g. $\quad y = |2x - 4|$

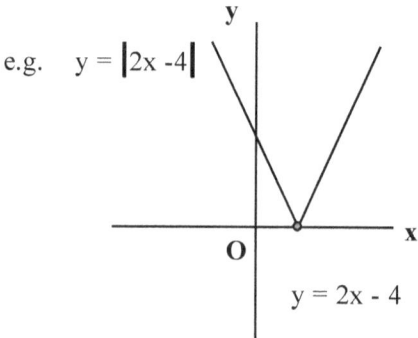

$y = 2x - 4$

e.g.

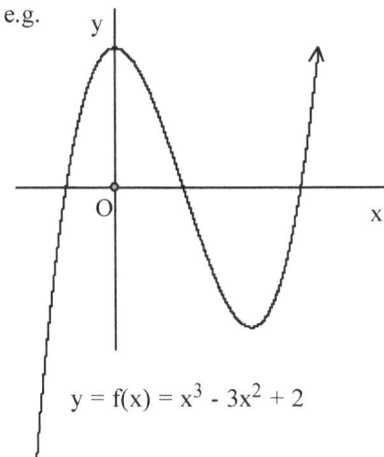

$y = f(x) = x^3 - 3x^2 + 2$

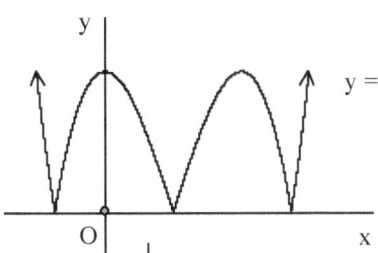

$y = |f(x)| = |x^3 - 3x^2 + 2|$

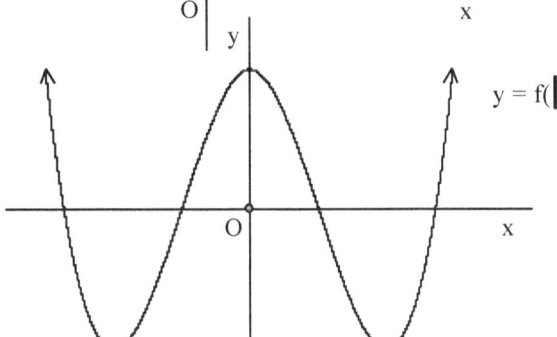

$y = f(|x|) = |x|^3 - 3|x|^2 + 2$

(5) Square Root Function

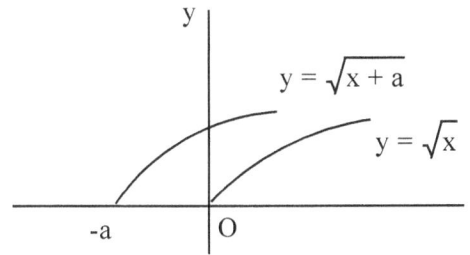

$f(x) = \sqrt{x}$

Domain: $[\, 0, \infty\,)$, Range: $[\, 0, \infty\,)$

$f(x) = \sqrt{x + a}$ $\qquad$ $a > 0$

Domain: $[\, -a\,, \infty\,)$, Range: $[\, 0, \infty\,)$

(6) Inverse Variations

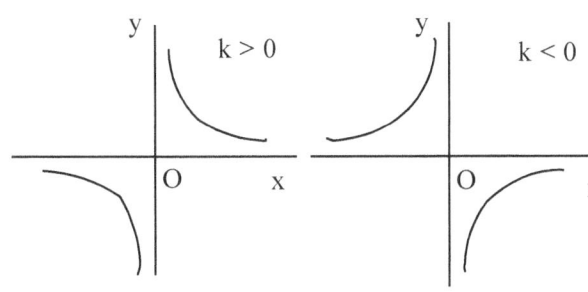

$xy = k$ $\quad$ or $\quad$ $y = \dfrac{k}{x}$

Horizontal Asymptote: $y = 0$
Vertical Asymptote: $x = 0$

To solve a problem:
$\qquad$ use $x_1 \cdot y_1 = x_2 \cdot y_2$

(7) Exponential Function

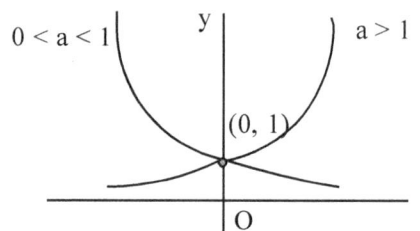

$y = a^x$ where a > 0 and a ≠ 1

    a > 1 , the function is increasing (exponential growth);
0 < a < 1 , the function is decreasing (exponential decay).

$y = (\dfrac{1}{a})^x = a^{-x}$ and $y = a^x$ are symmetric in the y-axis

Domain: {x | x all real numbers}

Range: {y | y > 0}

(8) Logarithmic Function

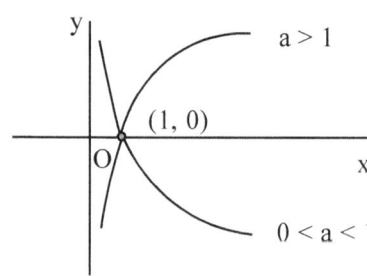

$y = \log_a x$ where a > 0 and a ≠ 1

$y = \log_a x$ is the inverse function of $y = a^x$

Domain: {x | x > 0 }

Range: {y | y all real numbers}

(9) Greatest Integer Function

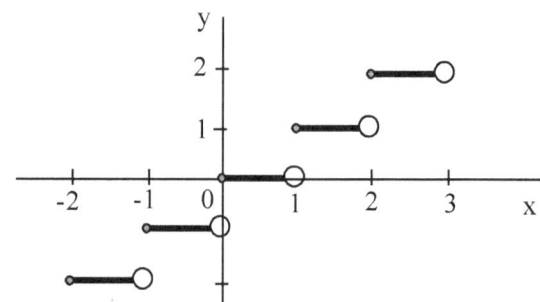

$y = [x]$

e.g. $y = [1.6] = 1$

Domain: {x | x all real numbers}

Range: {y | y all integers}

(10) Equation of a Circle (Relation, not Function)

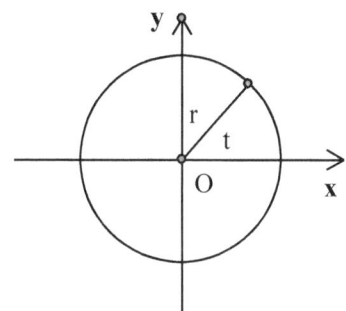

The equation of a circle with radius r and center at O(0, 0):

$$x^2 + y^2 = r^2$$

or $\quad y = \sqrt{r^2 - x^2} \quad$ and $\quad y = -\sqrt{r^2 - x^2}$

Parametric Equation:

$$x = r\cos t \ , \quad y = r\sin t \qquad 0 \le t \le 2\pi$$

(11) Ellipses, Circles (Relation, not Function)

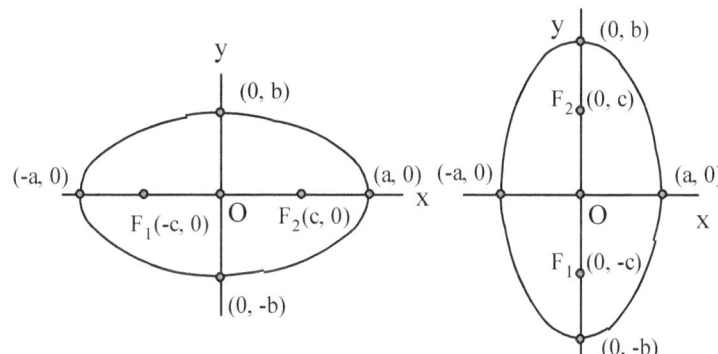

$$\frac{x^2}{a^2} + \frac{y^2}{b^2} = 1$$

It is a circle when a = b.

$$x^2 + y^2 = r^2$$

Parametric Equation:
$$x = a\cos t \ , \quad y = b\sin t \quad 0 \le t \le 2\pi$$

Area $= \pi ab$

$a > b > 0 \ , \quad c^2 = a^2 - b^2$
Major axis: 2a , Minor axis: 2b

$b > a > 0 \ , \quad c^2 = b^2 - a^2$
Major axis: 2b , Minor axis: 2a

(12) Hyperbolas (Relation, not Function)

$$\frac{x^2}{a^2} - \frac{y^2}{b^2} = 1$$

$y = -\frac{b}{a}x \qquad y = \frac{b}{a}x$

$F_1(-c, 0) \qquad F_2(c, 0)$

$(-a, 0) \qquad (a, 0)$

$c^2 = a^2 + b^2$
Foci: $F_1(-c, 0)$ and $F_2(c, 0)$

Asymptotes: $y = \pm \dfrac{b}{a}x$

$$\frac{y^2}{a^2} - \frac{x^2}{b^2} = 1$$

$y = -\frac{a}{b}x \qquad F_2(0, c) \qquad y = \frac{a}{b}x$

$(0, a)$

$(0, -a)$

$F_1(0, -c)$

$c^2 = a^2 + b^2$
Foci: $F_1(0, -c)$ and $F_2(0, c)$

Asymptotes: $y = \pm \dfrac{a}{b}x$

(13) Trigonometric Functions

y = sin x

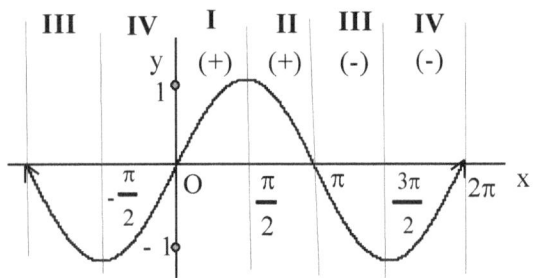

Domain: { x | x all real numbers }
Range: { y | - 1 ≤ y ≤ 1 }

y = cos x

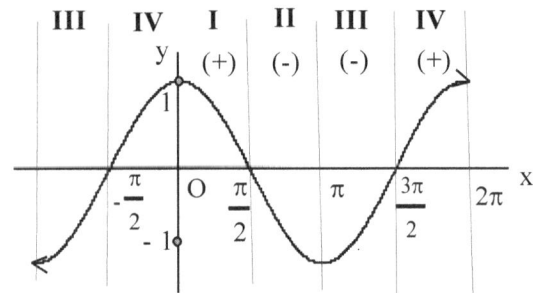

Domain: { x | x all real numbers }
Range: { y | - 1 ≤ y ≤ 1 }

Inverse Trigonometric Functions

y = arcsin x or y = sin⁻¹x

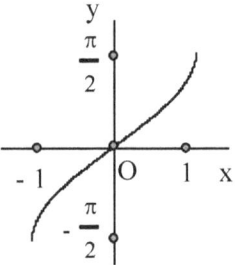

Domain: { x | - 1 ≤ x ≤ 1 }

Range : $\{ y \mid -\dfrac{\pi}{2} \le y \le \dfrac{\pi}{2} \}$

y = arccos x or y = cos⁻¹x

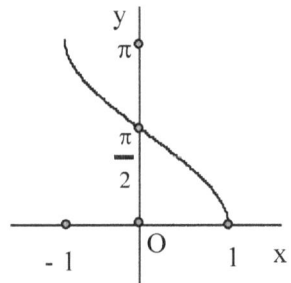

Domain: { x | - 1 ≤ x ≤ 1 }
Range : { y | 0 ≤ y ≤ π }

y = tan x

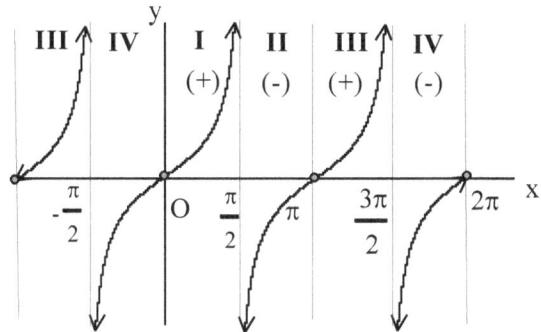

Domain: $\{ \, x \mid x \neq \dfrac{\pi}{2} + n\pi \text{ for n an integer} \, \}$

Range: $\{ \, y \mid y \text{ all real numbers} \, \}$

y = arctan x or y = tan⁻¹x

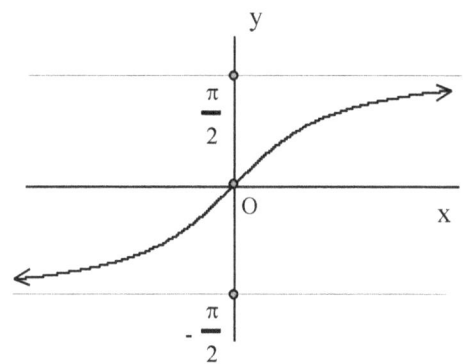

Domain: $\{ \, x \mid x \text{ all real numbers} \, \}$

Range: $\{ \, y \mid -\dfrac{\pi}{2} < y < \dfrac{\pi}{2} \, \}$

y = cot x

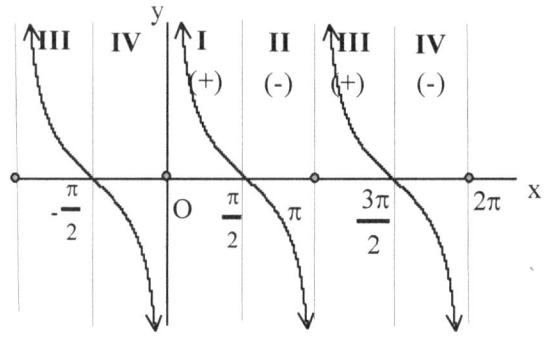

Domain: $\{ \, x \mid x \neq n\pi \text{ for n an integer} \, \}$
Range: $\{ \, y \mid y \text{ all real numbers} \, \}$

y = arccot x or y = cot⁻¹x

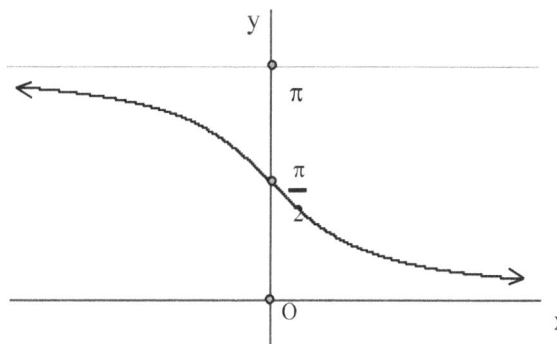

Domain: $\{ \, x \mid x \text{ all real numbers} \, \}$
Range: $\{ \, y \mid 0 < y < \pi \, \}$

II. Function Graphs

$$y = \sec x = \frac{1}{\cos x}$$

$$y = \text{arcsec } x = \sec^{-1}x = \cos^{-1}(\frac{1}{x})$$

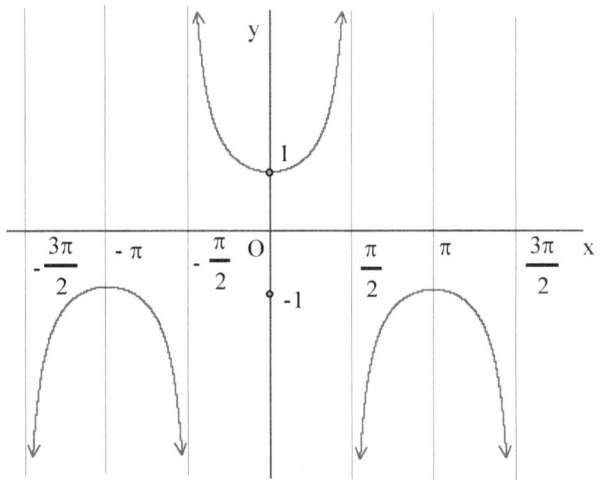

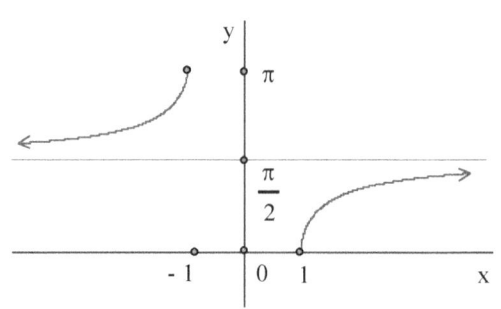

Domain: { x | x ≠ $\frac{\pi}{2}$ + nπ for n an integer }

Range: { y | (- ∞, -1] U [1, ∞) }

Domain: {x | x (- ∞, -1] U [1, ∞) }

Range: {y | 0 ≤ y < $\frac{\pi}{2}$, $\frac{\pi}{2}$ < y ≤ π }

$$y = \csc x = \frac{1}{\sin x}$$

$$y = \text{arccsc } x = \csc^{-1}x = \sin^{-1}(\frac{1}{x})$$

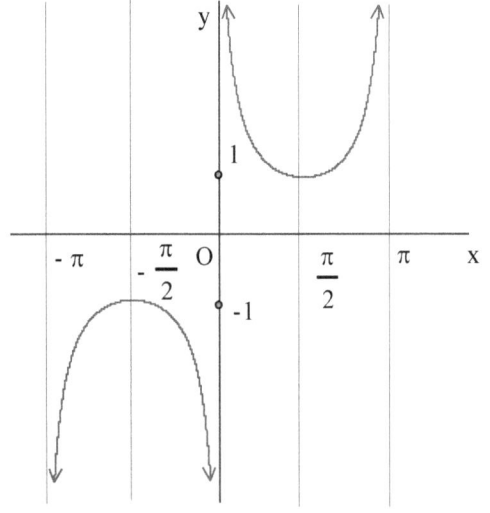

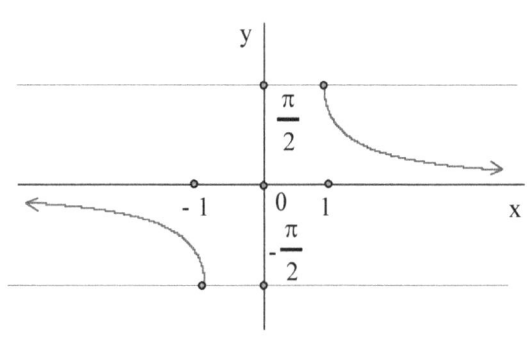

Domain: { x | x ≠ nπ for n an integer }

Range: { y | (- ∞, -1] U [1, ∞) }

Domain: {x | x (- ∞, -1] U [1, ∞) }

Range: {y | -$\frac{\pi}{2}$ ≤ y <0 , 0 < y ≤ $\frac{\pi}{2}$ }

2. Parametric Equations

If x and y are both given as functions of a third variable t, $x = x(t)$ and $y = y(t)$ are called parametric equations. The curve defined by these equations is called a parametric curve.

e.g.   The curve defined by $x = 2t$, $y = 5t - t^2$ is a parabola.

e.g.   The graph defined by $x = a + mt$, $y = b + nt$ is a line or a line segment through the point (a, b).

e.g.   The curve defined by $x = a\cos t$, $y = b\sin t$ ($0 \le t \le 2\pi$) is an ellipse.

Finding Cartesian Equations from Parametric Equations

e.g. $x = \sqrt{t}$, $y = 1 - t$, $t \ge 0$
$$x^2 = t$$
$$y = 1 - x^2, \qquad x \ge 0$$

e.g. $x = 3 + 5\cos t$, $y = 4 + 5\sin t$, $0 \le t \le 2\pi$
$$(x - 3)^2 = 5^2\cos^2 t , \ (y - 4)^2 = 5^2\sin^2 t$$
$$(x - 3)^2 + (y - 4)^2 = 5^2$$
It is a circle with radius 5 and center (3, 4).

e.g. $x = e^t$, $y = e^{-t}$
$$y = (e^t)^{-1} = x^{-1}$$
$$y = \frac{1}{x}, \ \text{ inverse variation with } x > 0$$

e.g. $x = \ln t$, $y = t^2$
$$e^x = t , \ \ y = (e^x)^2$$
$$y = e^{2x}$$

The Inverse Function in Parametric Forms

Function $y = f(x)$ in parametric forms: $x = g(t)$, $y = h(t)$.
Inverse Function $y = f^{-1}(x)$ in parametric forms: $x = h(t)$, $y = g(t)$.

Special Case: $y = f(x)$ can be written as $x = t$, $y = f(t)$; its inverse can be written as $x = f(t)$, $y = t$.

3. Polar Coordinates

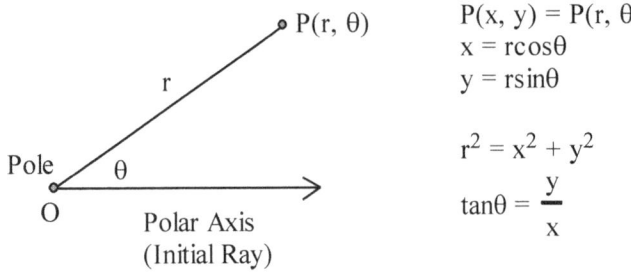

$$P(x, y) = P(r, \theta)$$
$$x = r\cos\theta$$
$$y = r\sin\theta$$

$$r^2 = x^2 + y^2$$
$$\tan\theta = \frac{y}{x}$$

(r, θ), $(-r, \theta + \pi)$, and $(-r, \theta - \pi)$ represent the same point. Therefore Polar coordinates are not unique.

(1) Line

$\theta = \alpha$ is a line through O making an angle α with the initial ray.

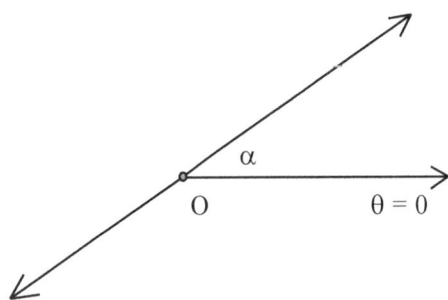

Polar coordinates can have negative r-values.

(2) Circles   radius = a

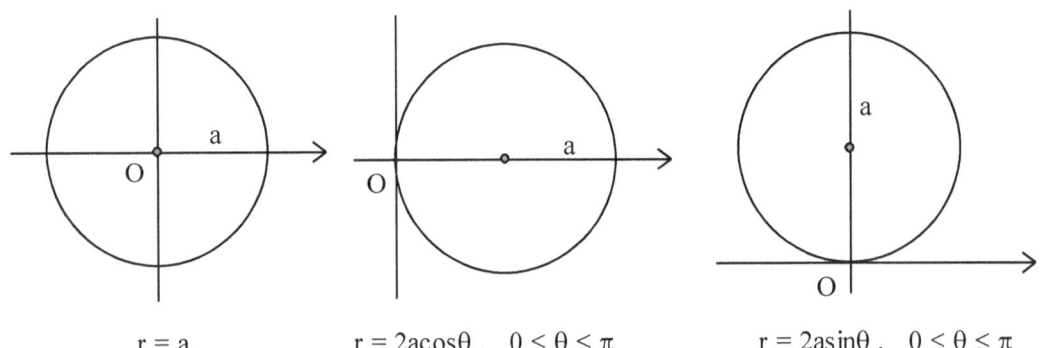

r = a   r = 2a\cos\theta , $0 \le \theta \le \pi$ r = 2a\sin\theta , $0 \le \theta \le \pi$

(3) Limaçon Curves

$$r = a + b\cos\theta \quad \text{and} \quad r = a + b\sin\theta$$

Cardioid Curves $\quad |a| = |b| \qquad a > 0, \quad 0 \le \theta \le 2\pi$

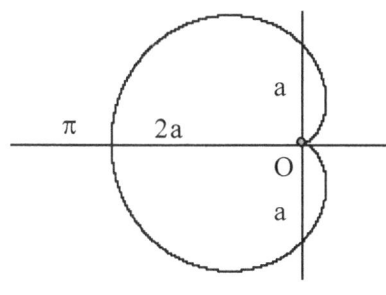

$$r = a(1 - \cos\theta)$$

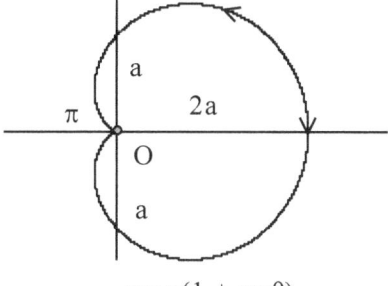

$$r = a(1 + \cos\theta)$$

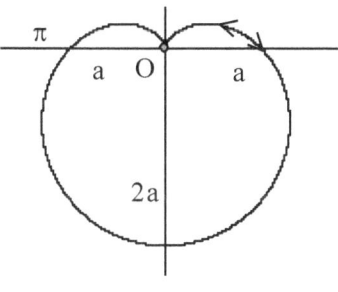

$$r = a(1 - \sin\theta)$$

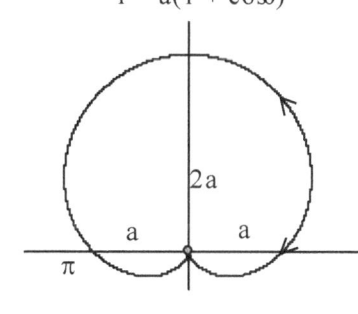

$$r = a(1 + \sin\theta)$$

Dimpled Limaçon $\quad |a| > |b| \qquad 0 \le \theta \le 2\pi$

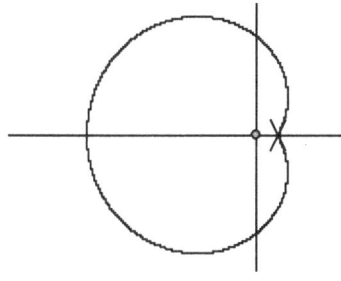

$$r = a - b\cos\theta$$

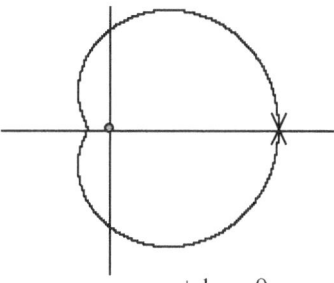

$$r = a + b\cos\theta$$

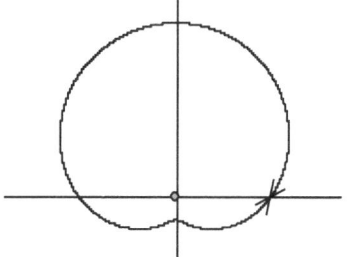

$$r = a - b\sin\theta$$

$$r = a + b\sin\theta$$

Limaçon with an Inner Loop $|a| < |b|$, $0 \leq \theta \leq 2\pi$

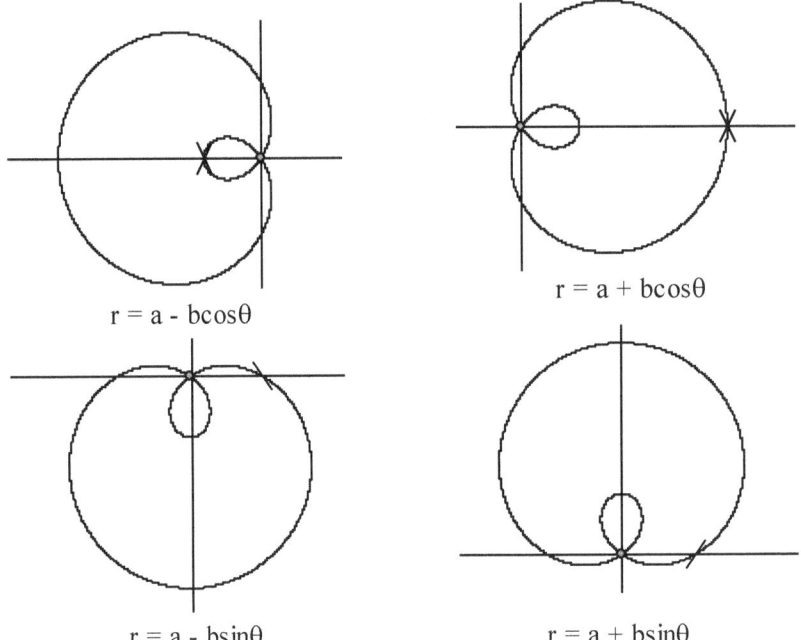

r = a - bcosθ

r = a + bcosθ

r = a - bsinθ

r = a + bsinθ

(4) Rose Curves

$$r = a\cos n\theta \text{ and } r = a\sin n\theta$$

when n is odd, the curve has n petals; when n is even, the curve has $2n$ petals.

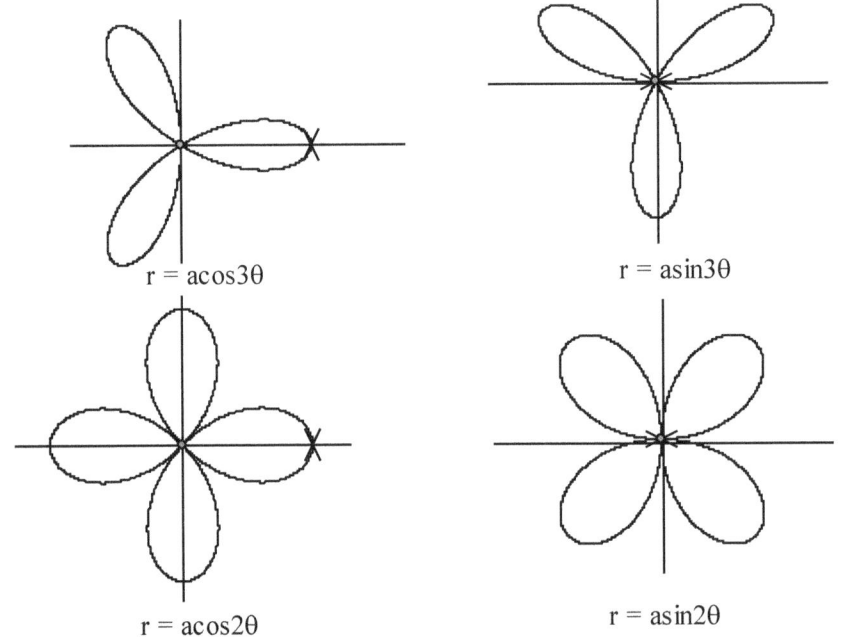

r = acos3θ

r = asin3θ

r = acos2θ

r = asin2θ

Symmetry of Polar Graphs

1. The polar graph is symmetric about the x-axis, if (r, θ) can be replaced by $(r, -\theta)$ or $(-r, \pi - \theta)$.

2. The polar graph is symmetric about the y-axis, if (r, θ) can be replaced by $(-r, -\theta)$ or $(r, \pi - \theta)$.

3. The polar graph is symmetric about the origin, if (r, θ) can be replaced by $(-r, \theta)$ or $(r, \theta + \pi)$.

e.g. Show that $r = 4\sin3\theta$ is symmetric about the y-axis.

$$(-r) = 4\sin(-3\theta) \qquad \text{replace } (r, \theta) \text{ by } (-r, -\theta)$$
$$-r = -4\sin3\theta$$
$$r = 4\sin3\theta$$

e.g. Determine the symmetries of the curve: $r^2 = 4\sin2\theta$.

(1) Replace (r, θ) by $(r, -\theta)$
$r^2 = 4\sin(-2\theta)$
$r^2 = -4\sin2\theta$ not the same, therefore it is not symmetric about the x-axis.

(2) Replace (r, θ) by $(-r, -\theta)$
$(-r)^2 = 4\sin(-2\theta)$
$r^2 = -4\sin2\theta$ not the same, therefore it is not symmetric about the y-axis.

(3) Replace (r, θ) by $(-r, \theta)$
$(-r)^2 = 4\sin2\theta$
$r^2 = 4\sin2\theta$ same as the original, therefore it is symmetric about the origin.

Converting from Polar to Rectangular

e.g. $r = 4\cos\theta$
$r^2 = 4r\cos\theta$
$x^2 + y^2 = 4x$
$x^2 - 4x + 4 + y^2 = 4$
$(x - 2)^2 + y^2 = 4$

Converting from Rectangular to Polar

e.g. $(x - 3)^2 + (y - 2)^2 = 13$
$x^2 - 6x + 9 + y^2 - 4y + 4 = 13$
$x^2 + y^2 - 6x - 4y = 0$
$r^2 - 6r\cos\theta - 4r\sin\theta = 0$
$r = 6\cos\theta + 4\sin\theta$

1. Limits

A function f(x) has a limit L as x approaches the number c (arbitrarily close, but not equal to c) if the graph of the function converges to one and only one point. Then we say that the limit of f(x) as x approaches c is L .

$$\lim_{x \to c} f(x) = L$$

If f(x) converges to a point as x approaches c from the left, then we say that f(x) has left-hand limit at c.
If f(x) converges to a point as x approaches c from the right, then we say that f(x) has right-hand limit at c.

A function f(x) has a limit at c if and only if the left-hand limit and the right-hand limit exist and are equal.

$$\lim_{x \to c^-} f(x) \;=\; \lim_{x \to c^+} f(x) \;=\; \lim_{x \to c} f(x)$$

x approaches to c, but not equal to c. The limit value of f(x) at c does not depend on f(c). Limit may exist at a point where the function is undefined.

Limits do not exist in the following situations:
1. There is a gap at the point x = c.
2. The function approaches infinity as x → c.
3. The function oscillates as x → c.

The basic functions that we have learned have limit at every point on their domains.
Arithmetic operation rules can be applied to the limits of functions.

2. Continuity

A function is continuous if it does not have breaks or holes.

(1) A function f(x) is continuous at point x = c, if

$$\lim_{x \to c} f(x) = f(c)$$

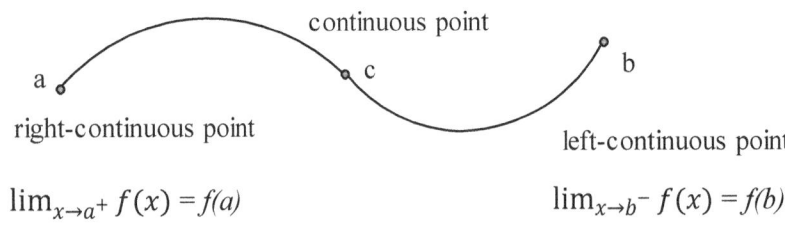

continuous point

a

right-continuous point

c

b

left-continuous point

$$\lim_{x \to a^+} f(x) = f(a) \qquad\qquad \lim_{x \to b^-} f(x) = f(b)$$

(2) The points of discontinuity are removable only if the function has limits at these points.

e.g.

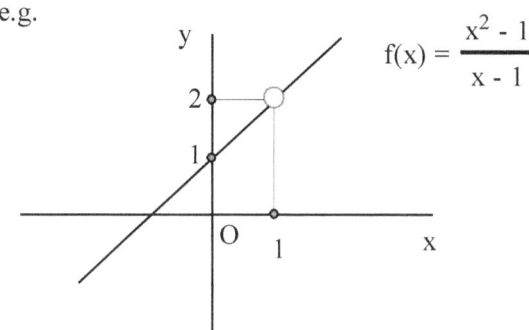

$$f(x) = \frac{x^2 - 1}{x - 1}$$

We can remove the discontinuity at x = 1 by define f (1) = 2 .

The basic functions that we have learned are continuous at every point on their domains. Continuous functions under algebraic operations are continuous where they are defined.

(3) Composition of Functions

f(g(x)) is continuous at x = c if g(x) is continuous at c and f(x) is continuous at g(c).

$$\lim_{x \to c} f(g(x)) = f(\lim_{x \to c} g(x)) = f(g(c))$$

The graphing calculator can help to determine limits , but not discontinuities.

e.g. $f(x) = \dfrac{\tan x}{x}$ on interval $(-\dfrac{\pi}{2}, \dfrac{\pi}{2})$

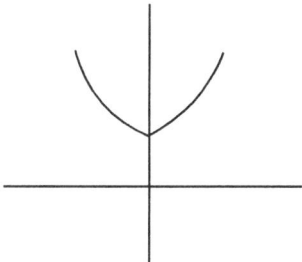

shown on graphing calculator

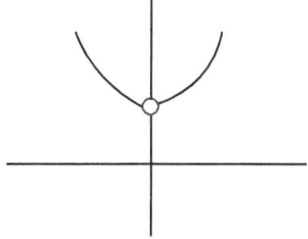

actual graph

e.g. A piecewise defined function

$$f(x) = \begin{cases} x^2 & \text{for } x \le 1 \\ -2x + 3 & \text{for } x > 1 \end{cases}$$

Show that f is continuous at x = 1.

$$\lim_{x \to 1^-} f(x) = \lim_{x \to 1^-} x^2 = 1$$

$$\lim_{x \to 1^+} f(x) = \lim_{x \to 1^+} -2x + 3 = 1$$

$$f(1) = 1$$

$$\lim_{x \to 1^-} f(x) = \lim_{x \to 1^+} f(x) = f(1),$$

therefore it is continuous at x = 1.

3. Infinitesimal and Infinity

A variable or a function approaches zero is called **infinitesimal**.
Infinitesimal is not a number. Its absolute vale can be smaller than any positive number except 0.
0 is not an infinitesimal. 0 is the limit of infinitesimal.

A variable or a function is called **infinity** if its absolute value can be greater than any number.
∞ or $+\infty$ is the symbol for positive infinity and $-\infty$ is the symbol for negative infinity.

Infinitesimal and infinity are reciprocals.

If $x \to +\infty$, then $\frac{1}{x} \to 0^+$ and if $x \to -\infty$, then $\frac{1}{x} \to 0^-$.

If $x \to 0^+$, then $\frac{1}{x} \to +\infty$ and if $x \to 0^-$, then $\frac{1}{x} \to -\infty$.

Note: $\lim_{x \to +\infty} \frac{1}{x} = 0$ and $\lim_{x \to -\infty} \frac{1}{x} = 0$, but

$\lim_{x \to 0} \frac{1}{x}$ does not exist , since $\lim_{x \to 0^+} \frac{1}{x} = +\infty$ and $\lim_{x \to 0^-} \frac{1}{x} = -\infty$

Note: $\lim_{x \to 0} \frac{1}{x^2} = \infty$

If the limit is infinity, we say it has infinite limit (sometimes we say it has no limit).

The infinitesimal and the infinity are dynamic variables, approaching in certain directions.

The Order of Infinitesimals and Infinities

If x is an infinitesimal, then x^2 is an infinitesimal of higher order.

Suppose that f(x) and g(x) are infinitesimals as $x \to c$, if

$\lim \frac{g}{f} = 0$,   then g is an infinitesimal of higher order to f, denoted as o(f);

$\lim \frac{g}{f} = \infty$,   then g is an infinitesimal of lower order to f;

$\lim \frac{g}{f} = c \neq 0$,   then g is an infinitesimal of same order to f;

$\lim \frac{g}{f} = 1$,   then g and f are equivalent infinitesimals, denoted as $g \sim f$.

Theorem: The difference between two equivalent infinitesimals is an infinitesimal of higher order.

$$g \sim f \quad \text{if and only if} \quad g = f + o(f)$$

e.g. $2x + 5x^2 - 3x^3 \sim 2x \quad$ as $\quad x \to 0 \quad$ (the infinitesimal of lowest order prevails)

The orders of infinities can be compared by the same way.

If x is an infinity ∞, then x^2 is an infinity of higher order, some books call it growing faster.

Suppose that f(x) and g(x) are infinities as $x \to c$, if

$\quad \lim \frac{g}{f} = \infty, \qquad$ then g is an infinity of higher order to f;

$\quad \lim \frac{g}{f} = 0, \qquad$ then g is an infinity of lower order to f;

$\quad \lim \frac{g}{f} = c \neq 0, \quad$ then g is an infinity of same order to f;

$\quad \lim \frac{g}{f} = 1, \qquad$ then g and f are equivalent infinities, denoted as $g \sim f$.

Theorem: The difference between two equivalent infinities is an infinity of lower order.

e.g $\quad 2x^3 + x^2 - 5x + 1 \sim 2x^3 \quad$ as $x \to \infty$ (the infinity of highest order prevails)

4. Finding Limits I.

Calculus is about finding limits.

(1) Plug in

$\quad \lim_{x \to c} f(x) = f(c) \qquad$ for all continuous functions.

e.g. $\quad \lim_{x \to 2} \frac{x^2 + 3x - 1}{2x} = \frac{(2)^2 + 3(2) - 1}{2(2)} = \frac{9}{4}$

$\quad \lim_{x \to 0} \frac{\cos x}{2 + x} = \frac{\cos 0}{2 + 0} = \frac{1}{2} \qquad \lim_{x \to 0} \frac{\sin x}{2 + x} = \frac{\sin 0}{2 + 0} = \frac{0}{2} = 0$

(2) Use Equivalent Infinitesimals

When $x \to 0$, the following infinitesimals are equivalent.

$$x \sim \sin x \sim \tan x \sim \sin^{-1} x \sim \tan^{-1} x \sim \ln(1 + x) \sim (e^x - 1)$$

The infinitesimal factors can be replaced by their equivalent ones in product or quotient.

e.g. $\quad \lim_{x \to 0} \frac{\tan 3x}{\sin 2x} = \lim_{x \to 0} \frac{3x}{2x} = \frac{3}{2}$

e.g. $\lim_{x\to 0}\dfrac{\sin 2x+\ln\sqrt{1+x}}{\tan x}=\lim_{x\to 0}\dfrac{\sin 2x}{\tan x}+\lim_{x\to 0}\dfrac{\ln\sqrt{1+x}}{\tan x}$

$=\lim_{x\to 0}\dfrac{\sin 2x}{\tan x}+\dfrac{1}{2}\lim_{x\to 0}\dfrac{\ln(1+x)}{\tan x}=\lim_{x\to 0}\dfrac{2x}{x}+\dfrac{1}{2}\lim_{x\to 0}\dfrac{x}{x}$

$=2+\dfrac{1}{2}=2\dfrac{1}{2}$

The infinitesimals can not be replaced by their equivalent ones in addition or subtraction.

e.g. $\lim_{x\to 0}\dfrac{\sin x-\tan x}{2x^3}\ne\lim_{x\to 0}\dfrac{x-x}{2x^3}=0$

(3) Use Equivalent Infinities

e.g. $\lim_{x\to\infty}\dfrac{2x^3-3x^2+1}{5x^3+1}=\lim_{x\to\infty}\dfrac{2x^3}{5x^3}=\dfrac{2}{5}$

e.g. $\lim_{x\to-\infty}\dfrac{2x^3-3x^2+1}{5x^2+1}=\lim_{x\to-\infty}\dfrac{2x^3}{5x^2}=\lim_{x\to-\infty}\dfrac{2x}{5}=-\infty$

(4) Use Substitution

e.g. $\lim_{x\to\infty}x\sin\dfrac{1}{x}=\lim_{x\to\infty}\dfrac{\sin\frac{1}{x}}{\frac{1}{x}}=\lim_{t\to 0^+}\dfrac{\sin t}{t}=1$ (let $t=\dfrac{1}{x}$)

(Hint: $\lim_{x\to 0}\dfrac{\sin x}{x}=1$ implies $\lim_{x\to 0^+}\dfrac{\sin x}{x}=1$ and $\lim_{x\to 0^-}\dfrac{\sin x}{x}=1$)

(5) Eliminate Subtraction

e.g. $\lim_{x\to 0}\dfrac{1-\cos x}{\sin x}=\lim_{x\to 0}\dfrac{1-\cos x}{\sin x}\cdot\dfrac{1+\cos x}{1+\cos x}=\lim_{x\to 0}\dfrac{\sin^2 x}{\sin x(1+\cos x)}$

$=\lim_{x\to 0}\dfrac{\sin x}{1+\cos x}=\dfrac{0}{2}=0$

e.g. $\lim_{x\to 0}\dfrac{1-\cos x}{x}=0$ (Hint: $x\sim\sin x$, use the result from the e.g. above.)

e.g. $\lim_{x\to 3}\dfrac{\sqrt{x+1}-2}{x-3}=\lim_{x\to 3}\dfrac{\sqrt{x+1}-2}{x-3}\cdot\dfrac{\sqrt{x+1}+2}{\sqrt{x+1}+2}=\lim_{x\to 3}\dfrac{x-3}{(x-3)(\sqrt{x+1}+2)}$

$=\lim_{x\to 3}\dfrac{1}{\sqrt{x+1}+2}=\dfrac{1}{4}$

e.g. $\lim_{x\to 2} \dfrac{\frac{4}{x}-2}{x-2} = \lim_{x\to 2} \dfrac{\frac{4-2x}{x}}{x-2} = \lim_{x\to 2} \dfrac{4-2x}{x(x-2)}$

$$= \lim_{x\to 2} \dfrac{-2(x-2)}{x(x-2)} = \lim_{x\to 2} \dfrac{-2}{x} = -1$$

(6) Composition of Functions

e.g. $\lim_{x\to\infty} \dfrac{\sqrt{x}}{\sqrt{2x+1}} = \lim_{x\to\infty} \sqrt{\dfrac{x}{2x+1}} = \sqrt{\lim_{x\to\infty} \dfrac{x}{2x+1}} = \sqrt{\dfrac{1}{2}} = \dfrac{\sqrt{2}}{2}$

(7) Some Important Limits

1. $\lim_{x\to 0} \dfrac{\sin x}{x} = 1$

2. $\lim_{x\to\infty} \left(1+\dfrac{1}{x}\right)^x = e$,　　　$\lim_{x\to 0}(1+x)^{\frac{1}{x}} = e$

    $\lim_{x\to\infty} \left(1+\dfrac{1}{x}\right)^{x+a} = e$　　　a is a constant

3. $\lim_{x\to -\infty} e^x = 0$,　　　　　$\lim_{x\to\infty} e^x = \infty$

    $\lim_{x\to -\infty} q^x = \infty$,　　　　$\lim_{x\to\infty} q^x = 0$　　　　$0 < q < 1$

4. $\lim_{x\to 0^+} \ln x = -\infty$,　　　$\lim_{x\to\infty} \ln x = \infty$

5. $\lim_{x\to\infty} \dfrac{\ln x}{x^n} = 0$,　　　　$\lim_{x\to\infty} \dfrac{x^n}{e^x} = 0$　　　n is a positive integer

    $\lim_{x\to\infty} \dfrac{\log_a x}{x^b} = 0$,　　　$\lim_{x\to\infty} \dfrac{x^b}{a^x} = 0$　　　$a > 1$ and $b > 0$

6. $\lim_{x\to\infty} \dfrac{x^b}{x^a} = \infty$　　　$a < b$

7. $\lim_{x\to 0} \dfrac{(1+x)^a - 1}{ax} = 1$　　　a is a real number

    $\lim_{x\to 0} \dfrac{(1+x)^a}{1+ax} = 1$

1. Derivative

The derivative is the limit of a ratio of two infinitesimals $\dfrac{dy}{dx}$.

The derivative is used as the rate of change $\dfrac{df}{dx}$, the slope $\dfrac{dy}{dx}$, and the $\tan\theta$.

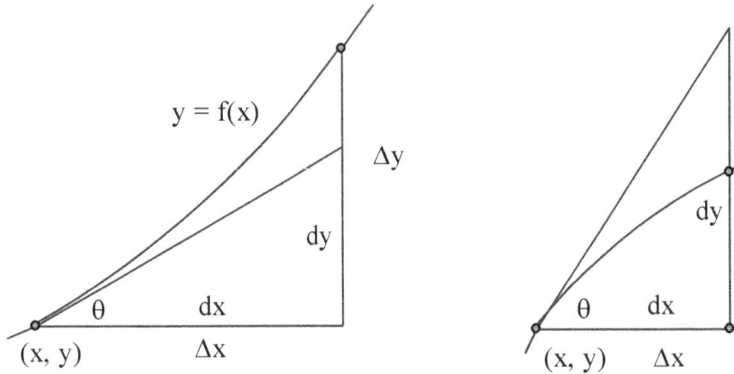

$\Delta x = dx$, $\Delta y \neq dy$ (Δy is the increment of f; dy is the leg of the right triangle in the diagram above.)

The differential dx is independent of x.
The differential dy is dependent on both x and dx.

Derivative forms: $f'(x) = \dfrac{dy}{dx} = \lim_{\Delta x \to 0} \dfrac{\Delta y}{\Delta x}$ Δx can be positive or negative

or $f'(x) = \lim_{\Delta x \to 0} \dfrac{f(x+\Delta x)-f(x)}{\Delta x}$

or $f'(x) = \lim_{h \to 0} \dfrac{f(x+h)-f(x)}{h}$

Other notations for derivative: $y'(x)$, y', f'.

e.g. $\dfrac{d}{dx}(3x^2) = (3x^2)'$

Differential form: $dy = \lim_{\Delta x \to 0} \Delta y = f'(x)\,dx$
or $df = f'(x)\,dx$

Second derivative: $f''(x) = \dfrac{d^2 y}{dx^2} = \dfrac{d}{dx}\left(f'(x)\right)$ $dx^2 = (dx)^2$

The nth derivative: $f^{(n)}(x) = \dfrac{d^n y}{dx^n}$ $dx^n = (dx)^n$

2. Differentiability

A function does not have derivatives at the points such as cusps, corners, or where the tangent is vertical.

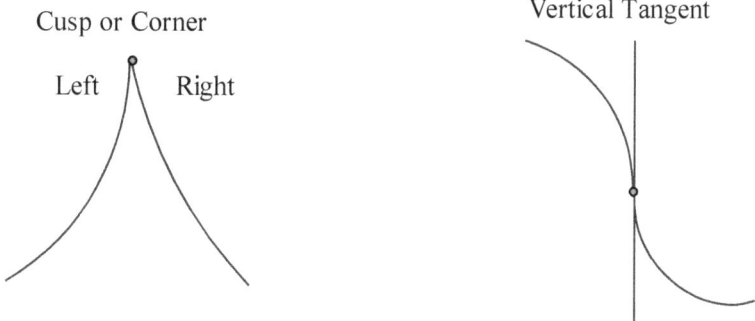

Cusp or Corner   Vertical Tangent

Left-hand derivative ≠ Right-hand derivative   Derivative undefined

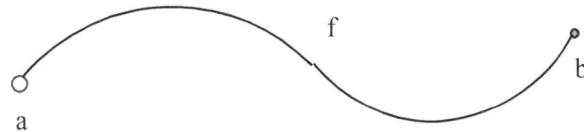

Function f has right-hand limit at point a , but f is discontinuous and not differentiable at point a.

Function f has left-hand limit at point b , and is left-hand continuous and left-hand differentiable at point b.

Differentiability implies continuity and continuity implies limit, but not vice versa.

The derivative of an odd function is an even function.
The derivative of an even function is an odd function.

e.g.   Determine the derivative at point x = 1 for function

$$f(x) = \begin{cases} x^2 & \text{for } x \le 1 \\ -2x + 3 & \text{for } x > 1 \end{cases}$$

$f'(x)|_{x = 1^-} = 2x|_{x = 1^-} = 2$ and $f'(x)|_{x = 1^+} = -2$

The left-hand derivative is not equal to the right -hand derivative.
Therefore f(x) is not differentiable at x = 1.

3. Derivative of Composite Functions

The Chain Rule: $\dfrac{dy}{dx} = \dfrac{dy}{du} \cdot \dfrac{du}{dx}$ where $y = y(u)$ and $u = u(x)$

e.g. $\dfrac{d}{dx}(x^2 + 1)^3 = 3(x^2 + 1)^2 \dfrac{d}{dx}(x^2 + 1) = 3(x^2 + 1)^2 \cdot 2x = 6x\,(x^2 + 1)^2$

            where $y = u^3$ and $u = (x^2 + 1)$

e.g. $\dfrac{d}{dx}(\sin^2 x) = 2\sin x \cdot \dfrac{d}{dx}(\sin x) = 2\sin x \cos x = \sin 2x$

            where $y = u^2$ and $u = \sin x$

4. Implicit Differentiation

Apply the Chain Rule to y.

e.g. $x^2 + y^2 = 25$

$\dfrac{d}{dx}(x^2) + \dfrac{d}{dx}(y^2) = \dfrac{d}{dx}(25)$

$2x + 2y\dfrac{dy}{dx} = 0$

$\dfrac{dy}{dx} = -\dfrac{2x}{2y} = -\dfrac{x}{y}$ y ' is expressed in terms of x and y

Its Second Derivative:

$y'' = \dfrac{d}{dx}\left(-\dfrac{x}{y}\right) = -\dfrac{y - x \cdot y'}{y^2} = -\dfrac{y - x \cdot (-\frac{x}{y})}{y^2} = -\dfrac{y^2 + x^2}{y^3}$

$= -\dfrac{25}{y^3}$

5. Derivative of Inverse Functions

The inverse function of $y = f(x)$ is $x = f(y)$ (implicit form)

or $y = f^{-1}(x)$ (explicit form)

The derivative of inverse function:

$\dfrac{dy}{dx} = \dfrac{1}{\dfrac{dx}{dy}}$

The derivative of inverse function at $x = a$:

$$\frac{dy}{dx}\Big|_{x=a} = \frac{1}{\dfrac{dx}{dy}\Big|_{y=f^{-1}(a)}}$$

or $$[f^{-1}(x)]'\big|_{x=a} = \frac{1}{f'(x)}\Big|_{x=f^{-1}(a)}$$

e.g. $f(x) = x^3 + 1$, find the derivative of $f^{-1}(x)$ at $x = 9$.

Rewrite the function as $y = x^3 + 1$.

Its inverse is $x = y^3 + 1$, when $x = 9$, $y = 2$

$$\frac{dy}{dx} = \frac{1}{\dfrac{dx}{dy}} = \frac{1}{3y^2} = \frac{1}{12}$$

e.g. Find $\dfrac{d}{dx}(\sin^{-1}x)$

$y = \sin x$, its inverse is $x = \sin y$ or $y = \sin^{-1}x$

$$\frac{dx}{dy} = \cos y = \sqrt{1 - \sin^2 y} = \sqrt{1 - x^2}, \qquad \cos y \geq 0 \text{ since the range of } y = \sin^{-1}x \text{ is } [-\frac{\pi}{2}, \frac{\pi}{2}]$$

$$\frac{d}{dx}(\sin^{-1}x) = \frac{1}{\dfrac{dx}{dy}} = \frac{1}{\sqrt{1 - x^2}}$$

Using Implicit Differentiation Method:

$x = \sin y$

$$\frac{d}{dx}(x) = \frac{d}{dx}(\sin y)$$

$$1 = \cos y \cdot \frac{dy}{dx}$$

$$\frac{dy}{dx} = \frac{1}{\cos y} = \frac{1}{\sqrt{1 - \sin^2 y}} = \frac{1}{\sqrt{1 - x^2}}$$

6. Derivative of Parametric Functions

$$\frac{dy}{dx} = \frac{\dfrac{dy}{dt}}{\dfrac{dx}{dt}} = \frac{y'(t)}{x'(t)}, \qquad \text{Second Derivative: } \frac{d}{dx}(f'(x)) = \frac{\dfrac{df'(x)}{dt}}{\dfrac{dx}{dt}}$$

e.g. Find the slpoe of the curve $x = 2\cos t$ and $y = 2\sin t$ at $t = \dfrac{\pi}{4}$.

$$\frac{dy}{dt} = \frac{d}{dt}(2\sin t) = 2\cos t$$

$$\frac{dx}{dt} = \frac{d}{dt}(2\cos t) = -2\sin t$$

$$\frac{dy}{dx} = \frac{\dfrac{dy}{dt}}{\dfrac{dx}{dt}} = \frac{2\cos t}{-2\sin t} = -\cot t$$

The slope at $t = \dfrac{\pi}{4}$ is

$$\frac{dy}{dx} = -\cot t \,\Big|_{t=\frac{\pi}{4}} = -\cot \frac{\pi}{4} = -1$$

e.g. Find the second derivative of the above function.

$$\frac{d}{dx}(-\cot t) = \frac{\dfrac{d}{dt}(-\cot t)}{\dfrac{dx}{dt}} = \frac{\csc^2 t}{-2\sin t} = -\frac{\csc^3 t}{2}$$

7. Derivative of Polar Functions

$r = r(\theta), \quad x = r \cdot \cos\theta = r(\theta) \cdot \cos\theta, \quad y = r \cdot \sin\theta = r(\theta) \cdot \sin\theta$

$$\frac{dy}{dx} = \frac{\frac{dy}{d\theta}}{\frac{dx}{d\theta}} = \frac{\frac{d}{d\theta}(r(\theta) \cdot \sin\theta)}{\frac{d}{d\theta}(r(\theta) \cdot \cos\theta)}$$

The rules are same as parametric functions.

Note: $\dfrac{dy}{dx}$ is the slope of the polar curve; $\dfrac{dr}{d\theta}$ is not the slope.

e.g. $r = 2\cos\theta$, write equations in terms of x and y for the tangent line and the normal

line to the curve at the point where $\theta = \dfrac{\pi}{6}$

$$\frac{dy}{dx} = \frac{\frac{dy}{d\theta}}{\frac{dx}{d\theta}} = \frac{\frac{d}{d\theta}(2\cos\theta \cdot \sin\theta)}{\frac{d}{d\theta}(2\cos\theta \cdot \cos\theta)}$$

$$= \frac{(\sin2\theta)'}{(1 + \cos2\theta)'} = \frac{\cos2\theta \cdot 2}{-\sin2\theta \cdot 2} = -\cot2\theta$$

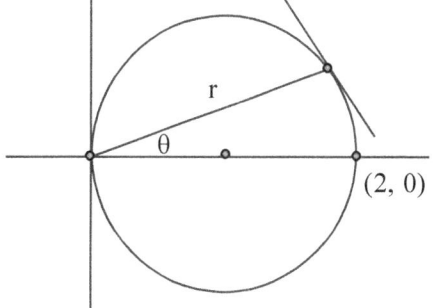

slope $m = \dfrac{dy}{dx} = -\cot2\theta = -\cot(2 \cdot \dfrac{\pi}{6}) = -\dfrac{\sqrt{3}}{3}$

tangent point: $x(\dfrac{\pi}{6}) = r \cdot \cos(\dfrac{\pi}{6}) = 2\cos(\dfrac{\pi}{6}) \cdot \cos(\dfrac{\pi}{6}) = \dfrac{3}{2}$

$y(\dfrac{\pi}{6}) = r \cdot \sin(\dfrac{\pi}{6}) = 2\cos(\dfrac{\pi}{6}) \cdot \sin(\dfrac{\pi}{6}) = \dfrac{\sqrt{3}}{2}$

The equation for the tangent line:

$$y - \frac{\sqrt{3}}{2} = -\frac{\sqrt{3}}{3}(x - \frac{3}{2})$$

The equation for the normal line:

the slope of the normal line $m_2 = \dfrac{-1}{m} = \sqrt{3}$

$$y - \frac{\sqrt{3}}{2} = \sqrt{3}(x - \frac{3}{2})$$

1. Curve Sketching

Graph of function f(x):

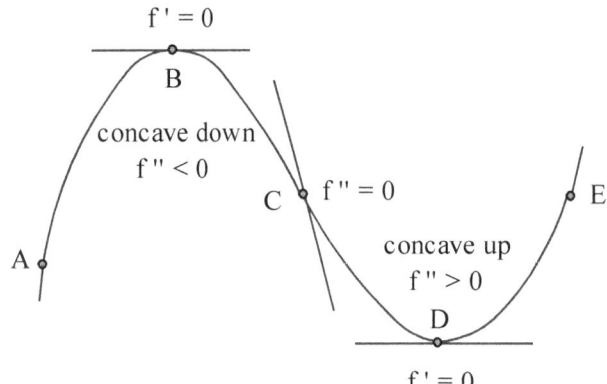

Point B is a point of local maximum (relative maximum).
Point D is a point of local minimum (relative minimum).
Point C is a point of inflection.

(1) f(x) is increasing where f'(x) > 0, A → B , D → E ;
     f(x) is decreasing where f'(x) < 0, B → C → D.

(2) f'(x) is increasing where f"(x) > 0, concave up, C → D → E ;
     f'(x) is decreasing where f"(x) < 0, concave down, A → B → C.

(3) Critical Points: f' = 0 or f' undefined , B and D.

(4) Local Extreme Values: f' changes signs at the point.
     Local maximum: f' changes from positive to negative, B;
     Local minimum: f' changes from negative to positive,   D.
     Local maximum values and local minimum values are called local extreme values.
     All possible extreme values are at critical points or end points, but not vice versa.

     e.g. $y = (x - 2)^4$, $y' = 4(x- 2)^3$
          $x \to 2^-$, y' < 0 ; $x \to 2^+$, y' > 0
          therefore y has a relative minimum at x = 2.

     e.g. $y = (x - 2)^3$, $y' = 3(x- 2)^2$ y' > 0 for all x except x = 2.
          x = 2 is a critical point but is not an extreme point.

(5) Points of Inflection: f" changes signs at the point
     All possible points of inflection are at f" = 0 or f" undefined, but not vice versa.

     e.g. $y = x^4$, $y" = 12x^2$
          y" = 0 at x = 0 , but it is not an inflection point, since y" > 0 for all x except x = 0.

Graph of f '(x):

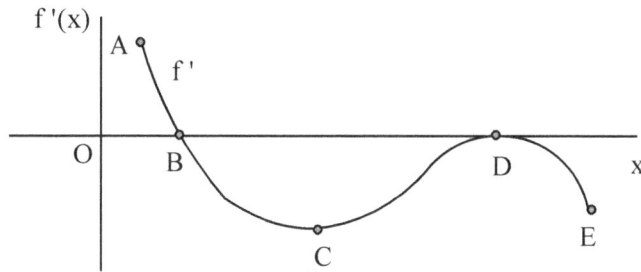

Point A: left endpoint, $f' > 0$, therefore $f(x)$ is increasing to the right.
  local minimum

Point B: $f' = 0$, f ' changes signs from positive to negative at the point.
  local maximum

Point C: $f'' = 0$, f '' changes signs (from negative to positive) at the point.
  point of inflection — an extreme of f '

Point D: $f' = 0$, f ' does not change sign at the point .
  not an extreme.

Point E: right endpoint, $f' < 0$, therefore $f(x)$ is decreasing from the left.
  local minimum

(1) f '(x) is decreasing where $f''(x) < 0$, concave down, $A \to B \to C$ and $D \to E$;
  f '(x) is increasing where $f''(x) > 0$, concave up, $C \to D$.

(2) Extremes of f ' are the points of inflection of f, C and D.

2. Maximum and Minimum Problems

Second Derivative Test:

Local maximum: $f' = 0$ and $f'' < 0$;
Local minimum: $f' = 0$ and $f'' > 0$

e.g. Find the local extreme values of $y = x^3 - 3x^2 + 2$.
  $y' = 3x^2 - 6x$, $y'' = 6x - 6$
  Solve $y' = 3x^2 - 6x = 0$, $x = 0$ and $x = 2$.
  When $x = 0$, $y'' = 6x - 6 = -6 < 0$, $y = x^3 - 3x^2 + 2 = 2$ is a local maximum ;
  When $x = 2$, $y'' = 6x - 6 = 6 > 0$, $y = x^3 - 3x^2 + 2 = -2$ is a local minimum.

V. Applications of Derivatives

Maximum Profit

r(x): the revenue from selling x items
c(x): the cost of producing x items
p(x): the profit from selling x items

$$p(x) = r(x) - c(x)$$

A production yields maximum profit when marginal revenue equals marginal cost.

$$\frac{dr}{dx} = \frac{dc}{dx} , \qquad \text{when } p'(x) = 0$$

3. Related Rates

A formula is an equation that relates two or more variables.
If variables change with time, then taking the derivative of the equation with respect to time will relate the rates of change of variables.

e.g.   Volume of a balloon: $V = \frac{4}{3}\pi r^3$ relates V and r

$$\frac{dV}{dt} = 4\pi r^2 \frac{dr}{dt}$$ relates $\frac{dV}{dt}$ and $\frac{dr}{dt}$

Using both equations to solve problems.

e.g.   In a right triangle: $x^2 + y^2 = 25$ relates leg x and leg y

$$2x\frac{dx}{dt} + 2y\frac{dy}{dt} = 0$$ relates $\frac{dx}{dt}$ and $\frac{dy}{dt}$

Using both equations to solve problems.

If the variable is decreasing, then the rate of change is negative.

4. Average Rate of Change

The average rate of change on the closed interval [a, b]

$$\frac{\Delta y}{\Delta x} = \frac{f(b) - f(a)}{b - a}$$

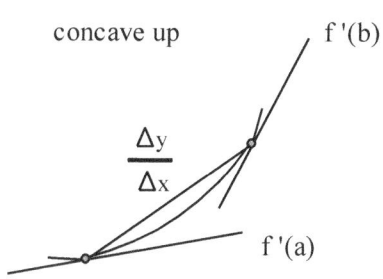

concave up   f '(b)

$\frac{\Delta y}{\Delta x}$

f '(a)

$\dfrac{\Delta y}{\Delta x}$ is the slope of the secant.

f '(a) and f '(b) are the slopes of the tangent lines.

If the curve is concave up, f ">0, then

$$f'(a) < \frac{\Delta y}{\Delta x} < f'(b) \qquad \text{since f' is increasing}$$

If the curve is concave down, f "<0, then

$$f'(a) > \frac{\Delta y}{\Delta x} > f'(b) \qquad \text{since f' is decreasing}$$

e.g.

t	0	3	5	8	10
f(t)	0	4	9	15	25

Use the data in the table to approximate the rate of change at t = 6.5.

$$f'(6.5) \approx \frac{f(8) - f(5)}{8 - 5} = \frac{15 - 9}{3} = 2$$

5. Linear Approximation

In local approximation we can use the differential dy = f '(a) dx to replace the increment $\Delta y = f(x) - f(a)$.

$$f(x) \approx f(a) + f'(a)(x - a)$$

If the curve is concave up, the answer is underestimated, since the tangent line is below the curve.
If the curve is concave down, the answer is overestimated, since the tangent line is above the curve.

e.g.   For any x near zero, $(1 + x)^k \approx 1 + kx$ for any number of k

e.g.   Estimate $\sqrt{4.2}$

$f(x) = \sqrt{x}$, $f'(x) = \dfrac{1}{2\sqrt{x}}$

$f(4) = \sqrt{4} = 2$, $f'(4) = \dfrac{1}{2\sqrt{4}} = \dfrac{1}{4}$

$f(x) \approx f(4) + f'(4)(x - 4) = 2 + \dfrac{1}{4}(x - 4)$

$f(4.2) \approx 2 + \dfrac{1}{4}(4.2 - 4) = 2 + \dfrac{1}{4}(0.2) = 2.05$

Since the curve is concave down, the answer is overestimated. $\sqrt{4.2} \approx 2.04939 < 2.05$

6. Finding Limits II.

(1) Use Derivative Definition

e.g. $\lim_{h \to 0} \dfrac{\sin\left(\frac{\pi}{2} + h\right) - \sin\left(\frac{\pi}{2}\right)}{h}$

$= \dfrac{d}{dx}(\sin x) \qquad$ at $x = \dfrac{\pi}{2}$

$= \cos x = \cos\dfrac{\pi}{2} = 0$

e.g. $\lim_{h \to 0} \dfrac{\sqrt{1+h} - 1}{h}$

$= \lim_{h \to 0} \dfrac{\sqrt{1+h} - \sqrt{1}}{h}$

$= \dfrac{d}{dx}(\sqrt{x}) \qquad$ at $x = 1$

$= \dfrac{1}{2\sqrt{x}} = \dfrac{1}{2}$

(2) Indeterminate Forms of Limits

$\dfrac{0}{0}$, $\dfrac{\infty}{\infty}$, $0 \cdot \infty$, $\infty - \infty$, 0^0, 1^∞, ∞^0

L' Hôpital's Rule:

If $\lim_{x \to a} f(x) = 0$, $\lim_{x \to a} g(x) = 0$, and $\lim_{x \to a} f'(x)$, $\lim_{x \to a} g'(x)$ exist, then

$$\lim_{x \to a} \dfrac{f(x)}{g(x)} = \lim_{x \to a} \dfrac{f'(x)}{g'(x)} \qquad\qquad \dfrac{0}{0}$$

It is also true for $x \to \pm\infty$.

L' Hôpital's Rule also applies to quotients that lead to the indeterminate form $\dfrac{\infty}{\infty}$.
It can be applied again if the result is $\dfrac{0}{0}$ or $\dfrac{\infty}{\infty}$.

e.g. $\lim_{x \to 0} \dfrac{e^x - 1}{x} = \lim_{x \to 0} \dfrac{e^x}{1} = 1$

e.g. $\lim_{x \to 0} \dfrac{\ln(1+x)}{x} = \lim_{x \to 0} \dfrac{\frac{1}{1+x}}{1} = 1$

e.g. $\lim_{x \to +\infty} \dfrac{\ln x}{x^n}$ $n > 0$, $\dfrac{\infty}{\infty}$

$= \lim_{x \to +\infty} \dfrac{\frac{1}{x}}{n\, x^{n-1}}$

$= \lim_{x \to +\infty} \dfrac{1}{n\, x^n} = 0$

The other indeterminate forms $0 \cdot \infty$, $\infty - \infty$, 0^0, 1^∞, ∞^0 can be transformed to either $\dfrac{0}{0}$ or $\dfrac{\infty}{\infty}$.

e.g. $\lim_{x \to 0}\left(\dfrac{1}{\sin x} - \dfrac{1}{x}\right)$ $\infty - \infty$

$= \lim_{x \to 0}\left(\dfrac{x - \sin x}{x \sin x}\right)$ $\dfrac{0}{0}$

$= \lim_{x \to 0}\left(\dfrac{1 - \cos x}{\sin x + x \cos x}\right)$ still $\dfrac{0}{0}$

$= \lim_{x \to 0}\left(\dfrac{\sin x}{\cos x + \cos x - x \sin x}\right)$

$= \dfrac{0}{2} = 0$

e.g. $\lim_{x \to 0^+} x^x$ 0^0

Let $f(x) = x^x$
$\ln f(x) = x \ln x$

$\lim_{x \to 0^+} \ln f(x) = \lim_{x \to 0^+} \dfrac{\ln x}{\frac{1}{x}}$ $\dfrac{\infty}{\infty}$

$= \lim_{x \to 0^+} \dfrac{\frac{1}{x}}{\frac{-1}{x^2}} = \lim_{x \to 0^+}(-x) = 0$

Therefore $\lim_{x \to 0^+} x^x = \lim_{x \to 0^+} f(x) = \lim_{x \to 0^+} e^{\ln f(x)} = e^0 = 1$

Note: If $\lim_{x \to a} \dfrac{f(x)}{g(x)}$ is not in indeterminate forms, L' Hôpital's Rule can not be applied.

e.g. $\lim_{x \to 5} \dfrac{x^2 + 1}{x} = \dfrac{5^2 + 1}{5} = \dfrac{26}{5}$

but $\lim_{x \to 5} \dfrac{(x^2 + 1)'}{x'} = \lim_{x \to 5} \dfrac{2x}{1} = 10$

VI. Antiderivatives

1. Antiderivative

$$\text{If} \quad \frac{d}{dx} F(x) = f(x), \qquad \text{then}$$

$$\int f(x)dx = F(x) + C \qquad \text{C is the arbitrary constant}$$

Integral sign $\int$ is the inverse operation of the differential sign d.
F(x) is called antiderivative, and the set of all antideravatives F(x) + C is called the indefinite integral.

The antiderivative can be written in the differential form:

$$\text{If} \quad dF(x) = f(x)dx, \qquad \text{then}$$

$$\int f(x)dx = F(x) + C \qquad \text{C is the arbitrary constant}$$

2. Rules of Integation

1. $\int c\, f(x)dx = c \int f(x)dx$
2. $\int [f(x) \pm g(x)]dx = \int f(x)dx \pm \int g(x)dx$
3. $\frac{d}{dx} \int f(x)dx = f(x) \qquad \text{or} \quad d \int f(x)dx = f(x)dx$
4. $\int \frac{d}{dx} f(x)dx = f(x) + C \quad \text{or} \quad \int df(x) = f(x) + C$

3. Substitution Rule

$$\text{If} \quad \int f(x)dx = F(x) + C, \qquad \text{then}$$

$$\int f(u)du = F(u) + C \qquad \text{where } u = u(x) \text{ is differentiable and } f(u) \text{ is continuous}$$

e.g. $\int \frac{dx}{1 + \sqrt{x-2}}$ $\qquad$ let $u = 1 + \sqrt{x-2}$, then $x = (u-1)^2 + 2$, $dx = 2(u-1)du$

$$= \int \frac{(2u-2)du}{u}$$

$$= \int 2du - \int \frac{2du}{u} = 2u - 2\ln|u| + C$$

$$= 2 + 2\sqrt{x-2} - 2\ln(1 + \sqrt{x-2}) + C$$

e.g. $\int x\sqrt{x-2}\,dx$ $\qquad$ let $u = \sqrt{x-2}$, then $x = u^2 + 2$, $dx = 2udu$

$$= \int (u^2 + 2)u \cdot 2udu$$

$$= \int (2u^4 + 4u^2)du$$

$$= \frac{2u^5}{5} + \frac{4u^3}{3} + c$$

$$= \frac{2}{5}(x-2)^{\frac{5}{2}} + \frac{4}{3}(x-2)^{\frac{3}{2}} + C$$

e.g. $\int x\sin(2x^2 + 1)dx$ let $u = 2x^2 + 1$, then $du = 4xdx$ or $xdx = \dfrac{du}{4}$

$$= \int \frac{\sin u\,du}{4} = \frac{-1}{4}\cos u + C$$

$$= \frac{-1}{4}\cos(2x^2 + 1) + C$$

e.g. $\int \dfrac{\sin\left(\frac{1}{x}\right)}{x^2}dx$ let $u = \dfrac{1}{x}$, then $x = \dfrac{1}{u}$, $dx = d\left(\dfrac{1}{u}\right) = -\dfrac{1}{u^2}du$

$$= \int \frac{\sin u}{\left(\frac{1}{u}\right)^2}\left(-\frac{1}{u^2}\right) du$$

$$= -\int \sin u\,du = \cos u + C = \cos\frac{1}{x} + C$$

Trigonometric Substitution

Let $x = a\sin t$ for $\sqrt{a^2 - x^2}$ $-\dfrac{\pi}{2} \le t \le \dfrac{\pi}{2}$

Let $x = a\sec t$ for $\sqrt{x^2 - a^2}$ $0 \le t < \dfrac{\pi}{2}$, $x \ge a$ or $\dfrac{\pi}{2} < t \le \pi$, $x \le -a$ $(a > 0)$

Let $x = a\tan t$ for $\sqrt{x^2 + a^2}$ $-\dfrac{\pi}{2} < t < \dfrac{\pi}{2}$

e.g. $\int \sqrt{1 - x^2}dx$ let $x = \sin t$, $dx = \cos t\,dt$

$$= \int \sqrt{1 - \sin^2 t} \cdot \cos t\,dt \qquad \sqrt{1 - \sin^2 t} = \cos t$$

$$= \int \cos^2 t\,dt$$

$$= \int \left(\frac{1}{2} + \frac{\cos 2t}{2}\right) dt \qquad \text{since } \cos^2 x = \frac{1 + \cos 2x}{2}$$

$$= \frac{1}{2}t + \int \left(\frac{\cos 2t}{4}\right) d2t$$

$$= \frac{1}{2}t + \frac{\sin 2t}{4} + C = \frac{1}{2}t + \frac{2\sin t\cos t}{4} + C$$

$$= \frac{1}{2}\sin^{-1}x + \frac{1}{2}x\sqrt{1 - x^2} + C \qquad (\cos t = \sqrt{1 - \sin^2 t}, \ -\frac{\pi}{2} \le t \le \frac{\pi}{2})$$

4. Using Differential Identities

$$dx = \frac{1}{a}d(ax), \qquad dx = \frac{1}{a}d(ax + b)$$

$$x^n dx = \frac{1}{n+1}dx^{n+1} \qquad (n \ne -1)$$

$$xdx = \frac{dx^2}{2}, \qquad \frac{dx}{x} = d\ln x, \qquad \frac{dx}{x^2} = -d\left(\frac{1}{x}\right), \qquad \frac{dx}{\sqrt{x}} = 2d\sqrt{x} \quad (x > 0)$$

$$e^x dx = de^x$$

$$\sin x dx = -d\cos x \ , \qquad \cos x dx = d\sin x$$

$$\sec^2 x dx = d\tan x \ , \qquad \frac{dx}{\cos^2 x} = d\tan x$$

$$\int f(ax)dx = \frac{1}{a}\int f(ax)d(ax)$$

$$\int f(ax+b)dx = \frac{1}{a}\int f(ax+b)d(ax+b)$$

e.g. $\qquad \int \sin ax \, dx = \frac{1}{a}\int \sin ax \, d(ax)$

$$= -\frac{\cos ax}{a} + C$$

e.g. $\qquad \int \sin ax \cdot \cos ax \, dx = \frac{1}{a}\int \sin ax \cdot \cos ax \, d(ax)$

$$= \frac{1}{a}\int \sin ax \, d(\sin ax)$$

$$= \frac{1}{a}\cdot\frac{\sin^2 ax}{2} + C = \frac{\sin^2 ax}{2a} + C$$

e.g. $\qquad \int \frac{\sin \sqrt{x} dx}{\sqrt{x}} = 2\int \sin\sqrt{x} \, d\sqrt{x}$

$$= -2\cos\sqrt{x} + C$$

e.g. $\qquad \int \frac{x dx}{\sqrt{x^2-1}} = \frac{1}{2}\int \frac{d(x^2-1)}{\sqrt{x^2-1}}$

$$= \frac{1}{2}\int 2 \, d\sqrt{x^2-1}$$

$$= \sqrt{x^2-1} + C$$

5. Using Trigonometric Identities

$$\sin 2x = 2\sin x\cos x$$

$$\sin^2 x = \frac{1-\cos 2x}{2} \ , \qquad \cos^2 x = \frac{1+\cos 2x}{2} \ , \qquad \sin^2 x + \cos^2 x = 1$$

e.g. $\qquad \int \sin^2 x \, dx$

$$= \int \frac{1-\cos 2x}{2} \, dx = \int \frac{1}{2} \, dx - \int \frac{\cos 2x}{2} \, dx$$

$$= \frac{x}{2} - \frac{1}{4}\int \cos 2x \, d(2x)$$

$$= \frac{x}{2} - \frac{\sin 2x}{4} + C$$

e.g. $\quad \int \sin^3 x\, dx$
$= -\int \sin^2 x\, d\cos x$
$= -\int (1 - \cos^2 x)\, d\cos x \qquad$ since $\sin^2 x = 1 - \cos^2 x$
$= -\int d\cos x + \int \cos^2 x\, d\cos x$
$= -\cos x + \dfrac{\cos^3 x}{3} + C$

6. Completing the Square

$x^2 + bx + c$
$= x^2 + bx + (\dfrac{b}{2})^2 - (\dfrac{b}{2})^2 + c$
$= (x + \dfrac{b}{2})^2 - (\dfrac{b}{2})^2 + c$

$ax^2 + bx + c$
$= a(x^2 + \dfrac{b}{a} x + \dfrac{c}{a})$

e.g. $\quad \int \dfrac{dx}{x^2 + 2x + 2} \qquad\qquad x^2 + 2x + 2 = (x+1)^2 + 1, \quad$ let $u = x + 1$
$= \int \dfrac{du}{u^2 + 1}$
$= \tan^{-1} u + C = \tan^{-1}(x+1) + C$

e.g. $\quad \int \dfrac{dx}{\sqrt{2x^2 + 4x + 10}}$
$= \dfrac{1}{\sqrt{2}} \int \dfrac{dx}{\sqrt{x^2 + 2x + 5}}$
$= \dfrac{1}{\sqrt{2}} \int \dfrac{dx}{\sqrt{(x+1)^2 + 2^2}}$
$= \dfrac{\sqrt{2}}{2} \ln\left| x + 1 + \sqrt{x^2 + 2x + 5} \right| + C$

7. Integration by Parts

$$\int u\, dv = uv - \int v\, du$$

or $\qquad\qquad \int u \cdot v'\, dx = uv - \int v \cdot u'\, dx$

Usually u is in forms of $\ln x$, $\sin^{-1} x$, $\cos^{-1} x$, $\tan^{-1} x$, x^n.
$\quad v'$ is in forms of $\sin x$, $\cos x$, e^{ax}.

e.g. $\qquad \int x \cdot e^x\, dx \qquad\qquad\qquad u = x, \quad v' = e^x$
$\quad = \int x\, de^x = x \cdot e^x - \int e^x\, dx$
$\quad = x \cdot e^x - e^x + C$

e.g. $\int x^2 \cdot \cos x \, dx$ $u = x^2, \quad v' = \cos x$
$= \int x^2 \, d\sin x = x^2 \cdot \sin x - \int \sin x \, dx^2$
$= x^2 \cdot \sin x - 2 \int x \cdot \sin x \, dx$
$= x^2 \cdot \sin x + 2 \int x \, d\cos x$ use the parts formula again
$= x^2 \cdot \sin x + 2(x \cdot \cos x - \int \cos x \, dx)$
$= x^2 \cdot \sin x + 2(x \cdot \cos x - \sin x) + C$

Special Case: $\int f(x) dx = x \cdot f(x) - \int x \, df(x)$

e.g. $\int \ln x \, dx$
$= x \cdot \ln x - \int x \, d\ln x$
$= x \cdot \ln x - \int x \, \dfrac{dx}{x}$
$= x \cdot \ln x - x + C$

8. Integration by Partial Fractions

$\dfrac{f(x)}{g(x)}$ is a proper fraction, if not, do the long division first.

We only discuss the simple cases here. The denominator g(x) is a product of $(ax + b)$, $(ax + b)^2$, $(x^2 + c)$, or $(x^2 + c)^2$.

We want to write $\dfrac{f(x)}{g(x)}$ as a sum of simple fractions in forms of

$$\dfrac{A}{ax + b}, \quad \dfrac{A}{(ax + b)^2}, \quad \dfrac{Bx + C}{(x^2 + c)}, \quad \dfrac{Bx + C}{(x^2 + c)^2}$$

(1) $(ax + b)$ in g(x) corresponds to $\dfrac{A}{ax + b}$.

(2) $(ax + b)^2$ in g(x) corresponds to $\dfrac{A}{ax + b} + \dfrac{A_1}{(ax + b)^2}$.

(3) $(x^2 + c)$ in g(x) corresponds to $\dfrac{Bx + C}{(x^2 + c)}$.

(4) $(x^2 + c)^2$ in g(x) corresponds to $\dfrac{Bx + C}{(x^2 + c)} + \dfrac{B_1 x + C_1}{(x^2 + c)^2}$

e.g. $\int \dfrac{x}{(x-1)(x^2 + 1)} \, dx$

Let $\dfrac{x}{(x-1)(x^2 + 1)} = \dfrac{A}{x - 1} + \dfrac{Bx + C}{(x^2 + 1)}$

$\dfrac{x}{(x-1)(x^2 + 1)} = \dfrac{Ax^2 + A + Bx^2 - Bx + Cx - C}{(x - 1)(x^2 + 1)}$

$$\frac{x}{(x-1)(x^2+1)} = \frac{(A+B)x^2 + (C-B)x + A - C}{(x-1)(x^2+1)}$$

Compare and determine the coefficients:

$$A + B = 0 \quad (1), \quad\quad C - B = 1 \quad (2), \quad\quad A - C = 0 \quad (3)$$

Eq. (1) + Eq. (2) $\quad\quad A + C = 1 \quad\quad (4)$
Eq. (3) + Eq. (4) $\quad\quad 2A = 1$

$$A = \frac{1}{2}, \quad B = -\frac{1}{2}, \quad C = \frac{1}{2}$$

$$\int \frac{x}{(x-1)(x^2+1)} dx = \frac{1}{2} \int \left(\frac{1}{x-1} + \frac{-x+1}{x^2+1}\right) dx$$

$$= \frac{1}{2} \int \frac{dx}{x-1} - \frac{1}{2} \int \frac{x\,dx}{x^2+1} + \frac{1}{2} \int \frac{dx}{x^2+1}$$

$$= \frac{1}{2} \ln|x-1| - \frac{1}{4} \int \frac{d(x^2+1)}{x^2+1} + \frac{1}{2} \tan^{-1}x$$

$$= \frac{1}{2} \ln|x-1| - \frac{1}{4} \ln|x^2+1| + \frac{1}{2} \tan^{-1}x + C$$

e.g. $\quad\quad \int \frac{x}{(x-1)^2} dx$

Let $\quad\quad \frac{x}{(x-1)^2} = \frac{A}{x-1} + \frac{B}{(x-1)^2}$

$$\frac{x}{(x-1)^2} = \frac{A(x-1) + B}{(x-1)^2}$$

$$\frac{x}{(x-1)^2} = \frac{Ax - A + B}{(x-1)^2}$$

Compare and determine the coefficients:

$$A = 1, \quad\quad -A + B = 0 \quad\quad B = 1$$

$$\int \frac{x}{(x-1)^2} dx = \int \frac{1}{x-1} dx + \int \frac{1}{(x-1)^2} dx$$

$$= \ln|x-1| - \frac{1}{x-1} + C$$

1. Definite Integral

A definite integral on a closed interval [a, b] is the region between the function f(x) and the x-axis.

This region represents the sum of the products of a function f(x) and the increments of the variable Δx over the closed interval [a, b]. For example, the distance is the product of the speed and the time interval.

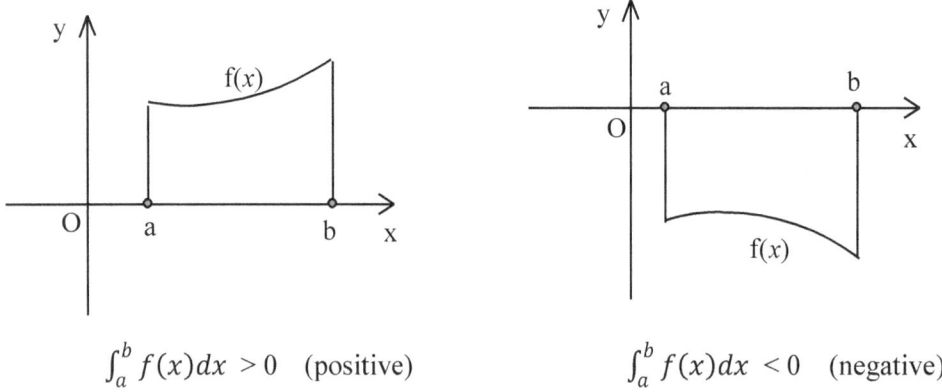

$\int_a^b f(x)dx > 0$ (positive) $\int_a^b f(x)dx < 0$ (negative)

 All continuous functions are integrable. That is, if a function is continuous on an interval [a, b], then its definite integral over [a, b] exists.

2. Calculating the Definite Integral

(1) Riemann Sum

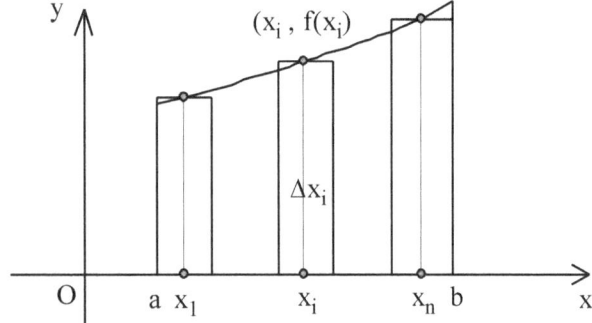

We divide the interval [a, b] into n subintervals Δx_i . x_i is any number in the corresponding subinterval Δx_i . The area of the rectangle on each subinterval is $f(x_i) \cdot \Delta x_i$. When the partitions become finer and finer, the sum of these areas approaches the real area between the curve of f(x) and the x-axis on the interval [a, b].

$$\int_a^b f(x)dx = \lim_{\max \Delta x_i \to 0} \sum_{i=1}^n f(x_i) \cdot \Delta x_i$$

f(x) is called the integrand;
a is the lower limit;
b is the upper limit.

If we divide the interval [a, b] into n equal subintervals Δx_i, then

$$\Delta x_i = \frac{b-a}{n} \quad \text{and} \quad x_i = a + i\Delta x_i = a + \frac{i}{n}(b-a)$$

$$\int_a^b f(x)dx \approx \sum_{i=1}^n f(a + \frac{i}{n}(b-a)) \cdot \frac{b-a}{n}$$

Riemann Sum for Approximation

Left Sum $L(n) = f(x_0) \cdot \Delta x_0 + f(x_1) \cdot \Delta x_1 + \cdots + f(x_{n-1}) \cdot \Delta x_{n-1}$

Right Sum $R(n) = f(x_1) \cdot \Delta x_1 + f(x_2) \cdot \Delta x_2 + \cdots + f(x_n) \cdot \Delta x_n$

Midpoint Sum $M(n) = f\left(\frac{x_0+x_1}{2}\right) \cdot \Delta x_1 + f\left(\frac{x_1+x_2}{2}\right) \cdot \Delta x_2 + \cdots + f\left(\frac{x_{n-1}+x_n}{2}\right) \cdot \Delta x_n$

Trapezoid Rule for Approximation

If the interval is divided by equal length $\Delta x = \frac{b-a}{n}$, then

$$T(n) = \left(\frac{y_0}{2} + y_1 + y_2 + \cdots + y_{n-1} + \frac{y_n}{2}\right)\Delta x$$

If the interval is not divided equally, then

$$T(n) = \frac{y_0+y_1}{2} \cdot \Delta x_1 + \frac{y_1+y_2}{2} \cdot \Delta x_2 + \cdots + \frac{y_{n-1}+y_n}{2} \cdot \Delta x_n$$

Here $y_0 = f_0 = f(x_0)$, $y_n = f_n = f(x_n)$.

e.g.

x	10	13	15	18	20
f(x)	19	23	29	35	45

Use trapezoid rule to find the average of f(x) on the interval [10, 20].

$$\text{Average} = \frac{1}{20-10} \int_{10}^{20} f(x)dx$$

$$\approx \frac{1}{10}\left(\frac{f_0+f_1}{2} \cdot \Delta x_1 + \frac{f_1+f_2}{2} \cdot \Delta x_2 + \cdots + \frac{f_{n-1}+f_n}{2} \Delta x_n\right)$$

$$= \frac{1}{10}\left[\frac{19+23}{2} \cdot (13\text{-}10) + \frac{23+29}{2} \cdot (15 - 13) + \frac{29+35}{2} \cdot (18 - 15) + \frac{35+45}{2}(20 - 18)\right]$$

$$= 29.1$$

(2) The Fundamental Theorem of Calculus, Part 1

$$\int_a^b f(x)dx = F(b) - F(a) \qquad \text{the net change of } F(x)$$

$f(x)$ is continuous and $F(x)$ is any antiderivative of $f(x)$ on [a, b].
This formula is also true for $b < a$.

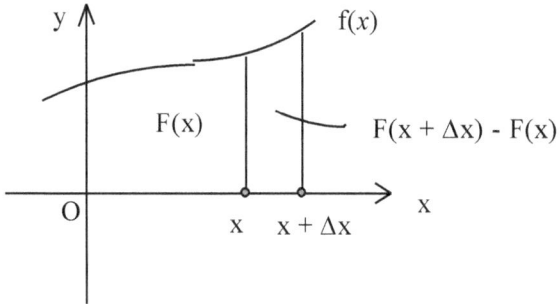

Let F(x) be the area of the region between f(x) and the x-axis.

$$F(x + \Delta x) - F(x) \approx f(x) \cdot \Delta x$$

$$\lim_{\Delta x \to 0} \frac{F(x + \Delta x) - F(x)}{\Delta x} = f(x)$$

That is
$$\frac{dF(x)}{dx} = f(x)$$

f(x) is the derivative of F(x) ; F(x) is any antiderivative of f(x).

3. Rules of Definite Integration

1. $\int_a^a f(x)dx = 0$
2. $\int_a^b cdx = c(b-a)$ where c is any constant
3. $\int_a^b cf(x)dx = c\int_a^b f(x)dx$ where c is any constant
4. $\int_a^b f(x)dx = -\int_b^a f(x)dx$
5. $\int_a^b f(x)dx = \int_a^c f(x)dx + \int_c^b f(x)dx$ (a, b, c can be in any order)
6. $\int_a^b [f(x) \pm g(x)]dx = \int_a^b f(x)dx \pm \int_a^b g(x)dx$
7. $\min f \cdot (b-a) \le \int_a^b f(x)dx \le \max f \cdot (b-a)$
8. If $f(x) \ge g(x)$ on [a, b] , then $\int_a^b f(x)dx \ge \int_a^b g(x)dx$.
   Special Case: If $f(x) \ge 0$ on [a, b] , then $\int_a^b f(x)dx \ge 0$.

4. Substitution Rule

$$\int_a^b f(u) \bullet u' dx = \int_{u(a)}^{u(b)} f(u) du$$

where $u'(x)$ is continuous on [a, b]

$f(u)$ is continuous on $[u(a), u(b)]$

e.g. $\int_1^4 \dfrac{\cos \sqrt{x}}{\sqrt{x}} dx$ let $t = \sqrt{x},\ x = t^2,\ dx = 2t dt$

$= \int_1^2 \dfrac{\cos t}{t} 2t dt$ $t = 1$ when $x = 1$

$t = 2$ when $x = 4$

$= 2 \int_1^2 \cos t\, dt$

$= 2\sin t \Big|_1^2 = 2\sin 2 - 2\sin 1$

e.g. $\int_0^{\frac{\pi}{2}} \sin^2 x \cdot \cos x dx$ let $t = \sin x,\quad dt = \cos x dx$

$= \int_0^1 t^2\, dt$ $t = 0$ when $x = 0$

$= \dfrac{t^3}{3} \Big|_0^1 = \dfrac{1}{3}$ $t = 1$ when $x = \dfrac{\pi}{2}$

5. Definite Integrals by Parts

$$\int_a^b u \cdot v' dx = uv \Big|_a^b - \int_a^b v \cdot u' dx$$

e.g. $\int_1^e x \ln x dx$

$= \dfrac{1}{2} \int_1^e \ln x dx^2$

$= \dfrac{1}{2} x^2 \cdot \ln x \Big|_1^e - \dfrac{1}{2} \int_1^e x^2\, d\ln x = \dfrac{1}{2} x^2 \cdot \ln x \Big|_1^e - \dfrac{1}{2} \int_1^e \dfrac{x^2 dx}{x}$

$= \dfrac{1}{2} x^2 \cdot \ln x \Big|_1^e - \dfrac{1}{2} \cdot \dfrac{x^2}{2} \Big|_1^e = \dfrac{e^2}{2} - \dfrac{e^2}{4} + \dfrac{1}{4}$

$= \dfrac{e^2}{4} + \dfrac{1}{4}$

6. Definite Integrals of Parametric Functions

Use Substitution Rule.

e.g.   Evaluate $\int_2^4 xy\,dx$, where $x = 2t$ and $y = e^t$.

$$\int_2^4 xy\,dx \qquad\qquad\qquad dx = 2dt$$

$$= \int_1^2 2t \cdot e^t \cdot 2dt \qquad\qquad t = 1 \text{ when } x = 2$$

$$= 4\int_1^2 t\, d\,e^t \qquad\qquad t = 2 \text{ when } x = 4$$

$$= 4\left[\, t \cdot e^t \,\Big|_1^2 - \int_1^2 e^t dt \,\right]$$

$$= 4\left[\, t \cdot e^t - e^t \,\right]_1^2 = 4e^2$$

7. The Fundamental Theorem of Calculus, Part 2

When a definite integral has a variable upper limit x, the integral itself defines a function $F(x)$, which is an antiderivative of the integrand $f(x)$.

If $F(x) = \int_a^x f(t)\,dt$,   then   where $f(t)$ is continuous on $[a, b]$

$F(x)$ is differentiable on $[a, b]$

$$\frac{dF(x)}{dx} = \frac{d}{dx}\int_a^x f(t)\,dt = f(x)$$

If $F(a) \neq 0$, then

$$F(x) = F(a) + \int_a^x f(t)\,dt$$

This formula is often used in problem solving.

e.g. The velocity of a particle is $v(t) = \dfrac{ds}{dt} = t \cdot e^t$ with initial condition $s = 5$ at $t = 1$.
Write an expression, involving an integral, for $s(t)$.

$$s(t) = s(1) + \int_1^t x \cdot e^x \,dx$$

e.g. $\dfrac{d}{dx}\int_a^x t \cdot \cos^2 t\,dt = x \cdot \cos^2 x$

Chain Rule is used when the upper limit is another function $g(x)$.

$$\frac{d}{dx}\int_a^{g(x)} f(t)dt = f(g(x)) \cdot \frac{dg(x)}{dx}$$

e.g. $\quad \dfrac{d}{dx} \displaystyle\int_a^{\sin x} t^2\, dt$

$$= \sin^2 x \cdot \dfrac{d(\sin x)}{dx}$$

$$= \sin^2 x \cdot \cos x$$

Both upper limit and lower limit are variables.

e.g. $\quad \dfrac{d}{dx} \displaystyle\int_x^{x^2} \sin^2 t\, dt \qquad\qquad\qquad$ (use $\int_a^b f(x)dx = \int_a^c f(x)dx + \int_c^b f(x)dx$)

$$= \dfrac{d}{dx}\Big[\ \int_x^1 \sin^2 t\, dt + \int_1^{x^2} \sin^2 t\, dt\ \Big] \quad \text{(c is any constant, we choose 1 here)}$$

$$= \dfrac{d}{dx}\int_1^{x^2} \sin^2 t\, dt - \dfrac{d}{dx}\int_1^{x} \sin^2 t\, dt \quad (\int_a^b f(x)dx = -\int_b^a f(x)dx)$$

$$= \sin^2(x^2) \cdot \dfrac{d}{dx}(x^2) - \sin^2 x$$

$$= 2x\sin^2(x^2) - \sin^2 x$$

8. Improper Definite Integrals

Type1: The interval of the integration is infinite

$$\int_a^\infty f(x)dx = \lim_{b\to\infty}\int_a^b f(x)dx$$

If the limit exists, the integral converges ; otherwise the integral diverges.

e.g. $\quad \displaystyle\int_0^\infty \dfrac{dx}{1+x^2}$

$$= \lim_{b\to\infty}\int_0^b \dfrac{dx}{1+x^2}$$

$$= \lim_{b\to\infty} \tan^{-1}b$$

$$= \dfrac{\pi}{2}$$

e.g. $\quad \displaystyle\int_{-\infty}^\infty \dfrac{dx}{1+x^2}$

$$= \int_{-\infty}^0 \dfrac{dx}{1+x^2} + \int_0^\infty \dfrac{dx}{1+x^2}$$

$$= 2\int_0^\infty \dfrac{dx}{1+x^2} \qquad\qquad \text{symmetric Property}$$

$$= \pi$$

e.g. $\int_1^\infty \frac{dx}{x}$

$$= \lim_{b \to \infty} \int_1^b \frac{dx}{x}$$

$$= \lim_{b \to \infty} \ln b = \infty$$

e.g. $\int_0^\infty \frac{x dx}{1 + x^2}$

$$= \lim_{b \to \infty} \frac{1}{2} \int_0^b \frac{d(1 + x^2)}{1 + x^2}$$

$$= \lim_{b \to \infty} \frac{1}{2} \ln(1 + b^2) = \infty$$

If the limit is infinite, the integral diverges.

e.g. $\int_0^\infty \frac{dx}{e^x}$

$$= \lim_{b \to \infty} \int_0^b \frac{dx}{e^x} = \lim_{b \to \infty} - \int_0^b e^{-x} d(-x)$$

$$= \lim_{b \to \infty} - e^{-x} \Big|_0^b = \lim_{b \to \infty} (e^0 - e^{-b}) = 1$$

e.g. $\int_0^\infty \cos x\, dx$

$$= \lim_{b \to \infty} \int_0^b \cos x\, dx$$

$$= \lim_{b \to \infty} \sin x \Big|_0^b = \lim_{b \to \infty} \sin b$$

The limit does not exist, the integral diverges.

Comparison Test

If $0 \le g(x) \le f(x)$ for all $x > a$, then

$\int_a^\infty g(x) dx$ converges if $\int_a^\infty f(x) dx$ converges,

or $\int_a^\infty f(x) dx$ diverges if $\int_a^\infty g(x) dx$ diverges.

Note: This Test is also true for Type 2 Improper Definite Integrals.

e.g. Compare $\int_1^\infty \frac{dx}{x^2}$ and $\int_1^\infty \frac{dx}{x^2+1}$

$$\int_1^\infty \frac{dx}{x^2} = \lim_{b\to\infty} \int_1^b \frac{dx}{x^2} = \lim_{b\to\infty} \left(-\frac{1}{x}\right) \bigg|_1^b = 1$$

Since $0 \le \frac{1}{x^2+1} \le \frac{1}{x^2}$ for all $x > 1$, $\int_1^\infty \frac{dx}{x^2+1}$ converges.

e.g. $\int_1^\infty \frac{\sin^2 x}{x^2} dx$ converges because $0 \le \frac{\sin^2 x}{x^2} \le \frac{1}{x^2}$ on $[1, \infty]$.

e.g. $\int_1^\infty \frac{dx}{\sqrt{x^2-0.1}}$ diverges because $\frac{1}{\sqrt{x^2-0.1}} \ge \frac{1}{x}$ on $[1, \infty]$.

e.g. $\int_0^\infty \frac{dx}{e^x+x}$ converges because $\frac{1}{e^x+x} \le \frac{1}{e^x}$ on $[0, \infty]$.

Limit Comparison Test

If $f(x)$ and $g(x)$ are both positive functions and

$$\lim_{x\to\infty} \frac{f(x)}{g(x)} = L \qquad\qquad L \text{ is a finite number, except } 0$$

then $\int_a^\infty f(x)dx$ and $\int_a^\infty g(x)dx$ both converge or both diverge.
(Infinitesimals of same order are either both convergent or both divergent.)

Note: $\int_a^\infty f(x)dx \ne L \cdot \int_a^\infty g(x)dx$

e.g. $\lim_{x\to\infty} \frac{1/x^2}{1/(x^2+1)} = 1$

but $\int_1^\infty \frac{dx}{x^2} = 1$ and $\int_1^\infty \frac{dx}{x^2+1} = \frac{\pi}{4}$

e.g. Compare $\int_1^\infty \frac{dx}{\sqrt{x}}$ and $\int_1^\infty \frac{dx}{\sqrt{2x+1}}$

$$\int_1^\infty \frac{dx}{\sqrt{x}} = \lim_{b\to\infty} \int_1^b \frac{dx}{\sqrt{x}}$$

$$= \lim_{b\to\infty} 2\sqrt{x} \bigg|_1^b$$

$$= \infty \qquad\qquad \text{diverges}$$

Do the limit comparison test:

$$\lim_{x\to\infty}\frac{f(x)}{g(x)} = \lim_{x\to\infty}\frac{\sqrt{2x+1}}{\sqrt{x}} = \lim_{x\to\infty}\sqrt{\frac{2x+1}{x}} = \sqrt{2}$$

Therefore $\int_1^\infty \frac{dx}{\sqrt{2x+1}}$ also diverges.

The following integral is often used for comparison:

$$\int_1^\infty \frac{dx}{x^p} = \frac{1}{p-1} \qquad\qquad \text{for } p > 1$$

$$\int_1^\infty \frac{dx}{x^p} \quad \text{diverges} \qquad\qquad \text{for } p \leq 1$$

$$\int_0^\infty \frac{dx}{e^x} = 1$$

$$|\sin x| \leq 1, \quad |\cos x| \leq 1$$

e.g. Determine the convergency of $\int_1^\infty \sin\left(\frac{1}{x}\right) dx$.

Use the limit comparison test:

$$\lim_{x\to\infty}\frac{\sin\left(\frac{1}{x}\right)}{\frac{1}{x}} = \lim_{u\to 0^+}\frac{\sin u}{u} = 1$$

$\int_1^\infty \frac{dx}{x}$ diverges, therefore $\int_1^\infty \sin\left(\frac{1}{x}\right) dx$ diverges.

Type 2: The function f(x) has infinite discontinuities in the interval.

$$\int_a^b f(x)dx = \lim_{x\to c}\int_a^b f(x)dx \qquad\qquad f(x) \to \infty \text{ when x} \to \text{c}$$

e.g. $\int_0^1 \frac{dx}{1-x}$ the integrand becomes undefined at $x = 1$

$$= \lim_{a\to 1^-}\int_0^a \frac{dx}{1-x} = \; = \lim_{a\to 1^-}[-\ln(1-x)]\Big|_0^a$$

$$= \lim_{a\to 1^-}[-\ln(1-a) + \ln(1-0)]$$

$$= \infty$$

e.g. $\int_0^1 \dfrac{dx}{\sqrt{1-x}}$ the integrand becomes undefined at $x = 1$

$$= \lim_{a \to 1^-} \int_0^a \dfrac{dx}{\sqrt{1-x}}$$

$$= \lim_{a \to 1^-} \left[-2\sqrt{1-x} \right]_0^a$$

$$= \lim_{a \to 1^-} \left[-2\sqrt{1-a} + 2\sqrt{1-0} \right]$$

$$= 2$$

e.g. $\int_0^1 \dfrac{dx}{\sqrt{1-x^2}}$ the integrand becomes undefined at $x = 1$

$$= \lim_{a \to 1^-} \int_0^a \dfrac{dx}{\sqrt{1-x^2}}$$

$$= \lim_{a \to 1^-} \sin^{-1} x \Big|_0^a = \lim_{a \to 1^-} \sin^{-1} a$$

$$= \dfrac{\pi}{2} \qquad \text{converges}$$

e.g. $\int_{-1}^1 \dfrac{dx}{x^2}$ the integrand becomes undefined at $x = 0$

$$= \int_{-1}^0 \dfrac{dx}{x^2} + \int_0^1 \dfrac{dx}{x^2}$$

$$= \lim_{a \to 0^-} \int_{-1}^a \dfrac{dx}{x^2} + \lim_{b \to 0^+} \int_b^1 \dfrac{dx}{x^2}$$

$$= \lim_{a \to 0^-} \left(-\dfrac{1}{x} \Big|_{-1}^a \right) + \lim_{b \to 0^+} \left(-\dfrac{1}{x} \Big|_b^1 \right)$$

$$= \lim_{a \to 0^-} \left(-\dfrac{1}{a} - 1 \right) + \lim_{b \to 0^+} \left(-1 + \dfrac{1}{b} \right)$$

$$= \infty \qquad \text{diverges}$$

In general, $\int_0^1 \dfrac{dx}{x^n}$ diverges when $n \geq 1$;
                         converges when $n < 1$.

Using the definite integral to find the product of two variables

e.g.   Area = b·h $\int y\,dx$,   Volume = B·h $\int A\,dx$

        Distance = v·t $\int v\,dt$ Work = F·s $\int F\,ds$

To set up a definite integral

1. First examine the curves of the functions and their boundaries, then determine the interval or intervals. Use symmetric property to simplify.

2. Divide the interval into subintervals and write a unique formula for an arbitrary partition $dF = f\,dx$, etc.

3. Integrate the formula over the interval $\int_a^b dF = \int_a^b f\,dx$, etc.

1. Area

In Rectangular Coordinates: $dA = y\,dx$ $(y > 0)$
                                $dA = x\,dy$ $(x > 0)$
        or more general: $dA = (y_2 - y_1)\,dx$ $(y_2 > y_1)$
                                $dA = (x_2 - x_1)\,dy$ $(x_2 > x_1)$

In Parametric Functions: $dA = y(t)\,dx(t)$ $(y > 0)$
                        or $dA = x(t)\,dy(t)$ $(x > 0)$

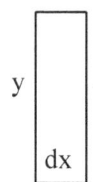

rectangular sector: dA = ydx

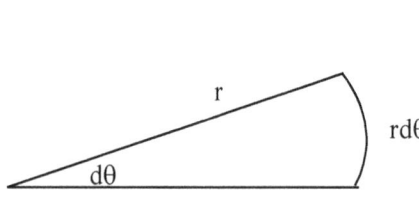

rectangular sector: dA = xdy

In Polar Coordinates:

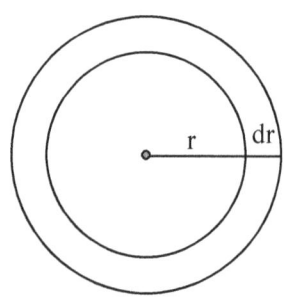

circular sector: $dA = \dfrac{1}{2} r^2\, d\theta$

ring sector: $dA = 2\pi\, r\,dr$

e.g. Find the area of the region between the x-axis and the graph of $y = \cos x$ on the interval $[0, \pi]$.

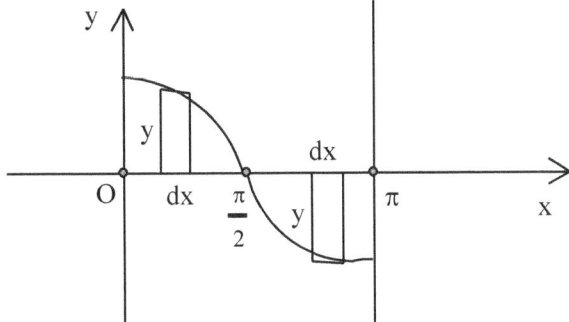

Partition $[0, \pi]$ with the zeros: $[\, 0, \frac{\pi}{2}\,]$ and $[\,\frac{\pi}{2}, \pi\,]$

$$\int_0^{\frac{\pi}{2}} \cos x\, dx = \sin x \,\Big|_0^{\frac{\pi}{2}} = \sin\frac{\pi}{2} - \sin 0 = 1$$

$$\int_{\frac{\pi}{2}}^{\pi} \cos x\, dx = \sin x \,\Big|_{\frac{\pi}{2}}^{\pi} = \sin \pi - \sin\frac{\pi}{2} = -1$$

Total enclosed area $A = 1 + |-1| = 2$.
(Note: Area is the absolute value of the definite integral.)

e.g. Find the area of the region between the curve $x = y^2 - 2y$ and the line $x = y$.

We denote x_2 for the right boundary $x_2 = y$ and x_1 for the left boundary $x_1 = y^2 - 2y$.

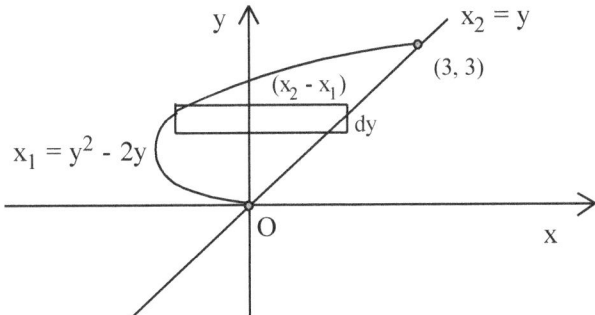

The formula for the area of the rectangular sector:

$$dA = (x_2 - x_1)dy = \left[\, y - (y^2 - 2y)\,\right]dy = \left(3y - y^2\right)dy$$

Integrate over the interval $[0, 3]$ of y :

$$A = \int_0^3 dA = \int_0^3 (3y - y^2)\,dy = \frac{3y^2}{2} - \frac{y^3}{3} \,\Big|_0^3 = \frac{27}{2} - \frac{27}{3} = \frac{27}{6} = \frac{9}{2}$$

e.g. Find the area enclosed by the cardioid $r = 2(1 - \cos\theta)$

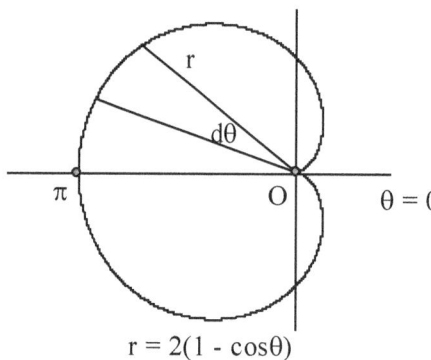

r = 2(1 - cosθ)

The formula for the area of the circular sector is

$$dA = \frac{1}{2}r^2 d\theta = \frac{1}{2} \cdot 2^2 (1 - \cos\theta)^2 d\theta$$

$$= 2(1 - 2\cos\theta + \cos^2\theta)d\theta$$

$$= 2(1 - 2\cos\theta + \frac{1}{2} + \frac{\cos 2\theta}{2}) \, d\theta$$

$$= (3 - 4\cos\theta + \cos 2\theta)d\theta$$

Integrate the circular sectors over the interval $[0, 2\pi]$.

$$A = \int_0^{2\pi} dA = \int_0^{2\pi} \frac{1}{2}r^2 d\theta = \int_0^{2\pi} (3 - 4\cos\theta + \cos 2\theta)d\theta$$

$$= \left[3\theta - 4\sin\theta + \frac{\sin 2\theta}{2} \right]_0^{2\pi}$$

$$= 6\pi$$

In general, cardioids $r = a(1 \pm \cos\theta)$ or $r = a(1 \pm \sin\theta)$

$$\text{Area bounded by curve} = 6\pi \left(\frac{a}{2}\right)^2$$

2. Volume

(1) Solids with Known Cross Section Area A(x) from x = a to x = b

$$V = \int_a^b A(x)\,dx$$

e.g. A solid has a circular base $x^2 + y^2 = 1$ in the xy-plane. The cross sections are squares perpendicular to the x-axis.

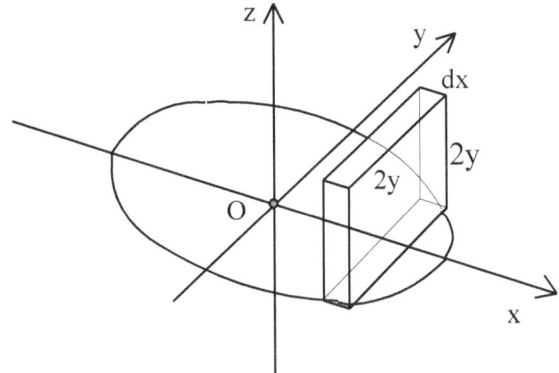

The area of cross section $A(x) = (2y)^2$ and the formula for the volume sector is

$$dV = A(x)\,dx = (2y)^2\,dx = 4(1 - x^2)\,dx$$

Integrate the volume sectors over the interval [-1, 1].

$$V = \int_{-1}^{1} dV = 4\int_{-1}^{1}(1 - x^2)\,dx$$

$$= 4\left[x - \frac{x^3}{3}\right]_{-1}^{1}$$

$$= \frac{16}{3}$$

(2) Solids of Revolution

The graph of the function revolved around x-axis:

Disk Sector: $dV = \pi r^2 dx$
Washer Sector: $dV = (\pi R^2 - \pi r^2)\, dx$ outer disk – inner disk
Shell Sector: $dV = 2\pi r \cdot h\, dy$ h: height of shell

e.g. The region bounded by the curve $y = \sqrt{x}$, $x = 4$ and the x-axis is revolved around the x-axis.

Method 1. Using Disk Sector: (strip perpendicular to the revolving axis)

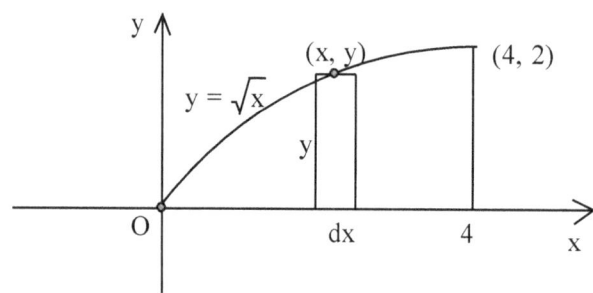

$$dV = \pi r^2 dx = \pi y^2 dx = \pi x\, dx$$

$$V = \int_0^4 dV = \int_0^4 \pi x\, dx = \left.\frac{\pi x^2}{2}\right|_0^4 = 8\pi$$

Method 2. Using the Cylindrical Shell Sector: (strip parallel to the revolving axis)

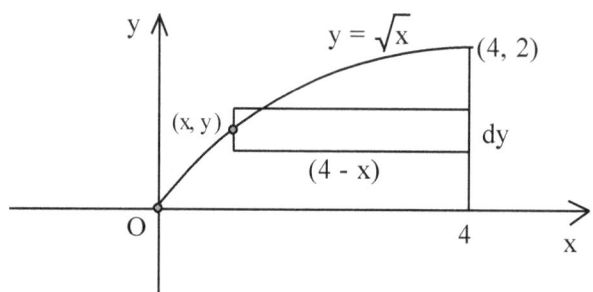

$$dV = 2\pi r \cdot h\, dy = 2\pi y(4 - x)dy = 2\pi y(4 - y^2)dy = (8\pi y - 2\pi y^3)dy$$

$$V = \int_0^2 dV = \int_0^2 (8\pi y - 2\pi y^3)dy = \left[\frac{8\pi y^2}{2} - \frac{2\pi y^4}{4}\right]_0^2 = 8\pi$$

e.g. The region bounded by the curve $y = \sqrt{x}$, $x = 4$, and the x-axis is rotated about the horizontal line y = 4.

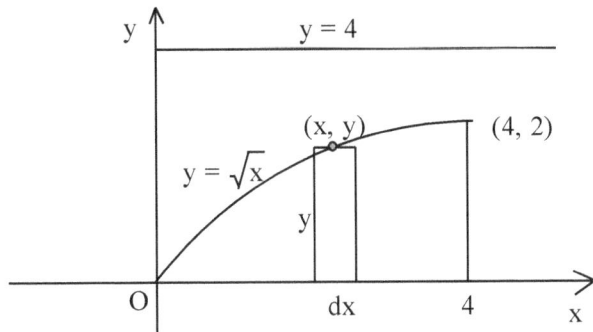

Using Washer Sector: (strip perpendicular to the revolving axis)

$$dV = (\pi R^2 - \pi r^2)dx = (\pi 4^2 - \pi(4-y)^2)dx = \left(\pi 4^2 - \pi(4-\sqrt{x})^2\right)dx$$

$$\begin{aligned}
V = \int_0^4 dV &= \int_0^4 \left(\pi 4^2 - \pi(4-\sqrt{x})^2\right)dx \\
&= \int_0^4 \left(16\pi - \pi(16 - 8\sqrt{x} + x)\right)dx \\
&= \int_0^4 \left(8\pi\sqrt{x} - \pi x\right)dx \\
&= \left[\frac{16\pi}{3} x^{\frac{3}{2}} - \frac{\pi}{2} x^2\right]_0^4 = \frac{104}{3}\pi
\end{aligned}$$

e.g. The region bounded by the curve $y = \sqrt{x}$, $= 2$, and the y-axis is revolved around the x-axis.

Using Washer Sector: (strip perpendicular to the revolving axis)

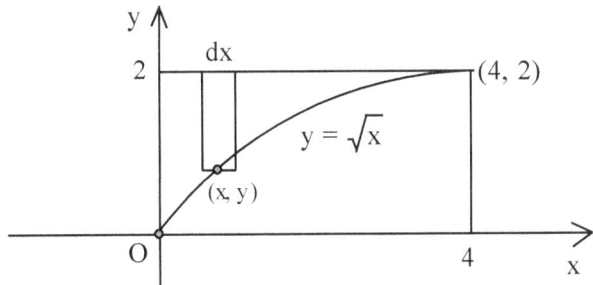

We denote y_2 for upper boundary $y_2 = 2$ and denote y_1 for lower boundary $y_1 = \sqrt{x}$.

$$dV = (\pi R^2 - \pi r^2)dx = (\pi y_2{}^2 - \pi y_1{}^2)dx = (4\pi - \pi x)dx$$

$$V = \int_0^4 dV = \int_0^4 (4\pi - \pi x)dx = \left[4\pi x - \frac{\pi x^2}{2}\right]_0^4 = 8\pi$$

(This problem can also be solved by using the cylindrical shell sector.)

3. Arc Length

We combine the Pythagorean Theorem and the limit concept to find the arc length.

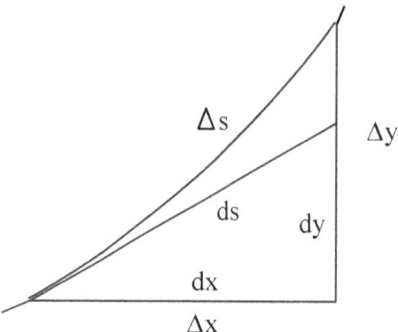

The arc length sector:

$$ds = \lim_{\Delta x \to 0} \Delta s = \lim_{\Delta x \to 0} \sqrt{(\Delta x)^2 + (\Delta y)^2} = \sqrt{(dx)^2 + (dy)^2}$$

In derivative form:

$$ds = \sqrt{1 + \left(\frac{dy}{dx}\right)^2} \, dx = \sqrt{1 + \left(\frac{dx}{dy}\right)^2} \, dy$$

Parametric Function: $x = x(t) \, , \, y = y(t)$

$$ds = \sqrt{\left(\frac{dx}{dt}\right)^2 + \left(\frac{dy}{dt}\right)^2} \, dt$$

In Polar Coordinates: $r = r(\theta)$

$$dr = \frac{dr}{d\theta} \cdot d\theta \, , \qquad (dr)^2 = \left(\frac{dr}{d\theta}\right)^2 \cdot (d\theta)^2$$

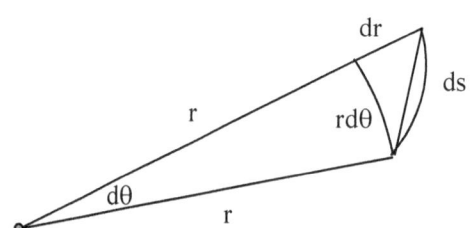

$$ds = \sqrt{(rd\theta)^2 + (dr)^2} = \sqrt{r^2(d\theta)^2 + (dr)^2}$$

$$ds = \sqrt{r^2(d\theta)^2 + \left(\frac{dr}{d\theta}\right)^2 (d\theta)^2}$$

$$= \sqrt{r^2 + \left(\frac{dr}{d\theta}\right)^2}\, d\theta$$

e.g. Find the arc length of the circle $x^2 + y^2 = 1$ in the first quadrant.

Method 1. In Rectangular Coordinates

Implicit Derivative: $2x + 2y\frac{dy}{dx} = 0$, $\frac{dy}{dx} = -\frac{x}{y}$

$$ds = \sqrt{1 + \left(\frac{dy}{dx}\right)^2}\, dx = \sqrt{1 + \frac{x^2}{y^2}}\, dx = \sqrt{\frac{y^2+x^2}{y^2}}\, dx$$

$$= \frac{dx}{\sqrt{1-x^2}}$$

$$S = \int_0^1 ds = \int_0^1 \frac{dx}{\sqrt{1-x^2}} = \sin^{-1}x \Big|_0^1 = \frac{\pi}{2}$$

Method 2. Use Parametric Function: $x = \cos t$, $y = \sin t$

$$ds = \sqrt{\left(\frac{dx}{dt}\right)^2 + \left(\frac{dy}{dt}\right)^2}\, dt = \sqrt{\sin^2 t + \cos^2 t}\; dt = dt$$

$$S = \int_0^{\frac{\pi}{2}} dt = \frac{\pi}{2}$$

Method 3. In Polar Coordinates

$$r = 1 , \qquad ds = \sqrt{r^2 + \left(\frac{dr}{d\theta}\right)^2}\, d\theta = \sqrt{1 + 0}\, d\theta$$

$$S = \int_0^{\frac{\pi}{2}} \sqrt{1 + 0}\, d\theta = \int_0^{\frac{\pi}{2}} d\theta = \frac{\pi}{2}$$

X. Applications of Integration

e.g.   Find the arc length of the cardioid $r = 2(1 - \cos\theta)$.

$$\frac{dr}{d\theta} = \frac{d}{d\theta}(2 - 2\cos\theta) = 2\sin\theta$$

$$ds = \sqrt{r^2 + \left(\frac{dr}{d\theta}\right)^2}\, d\theta = \sqrt{4(1 - \cos\theta)^2 + (2\sin\theta)^2}\, d\theta$$

$$= \sqrt{4(1 - 2\cos\theta + \cos^2\theta) + 4\sin^2\theta}\, d\theta$$

$$= \sqrt{8(1 - \cos\theta)}\, d\theta$$

$$= \sqrt{16 \cdot \sin^2\frac{\theta}{2}}\, d\theta$$

$$= 4\sin\frac{\theta}{2}\, d\theta$$

$$S = \int_0^{2\pi} ds = \int_0^{2\pi} 4\sin\frac{\theta}{2}\, d\theta = \int_0^{\pi} 8\sin t\, dt \qquad \text{let } t = \frac{\theta}{2}$$

$$= -8\cos t \Big|_0^{\pi} = -8(-1 - 1) = 16$$

In general, cardioids $r = a(1 \pm \cos\theta)$ or $r = a(1 \pm \sin\theta)$

Arc length of curve $= 8a$

4. Surface Area of Revolution

The surface area of a frustum of a cone:

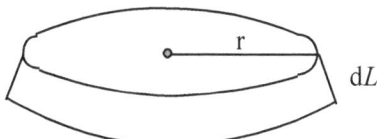

Band Sector: $dA = 2\pi r \, dL$

where $dL = \sqrt{(dx)^2 + (dy)^2} = \sqrt{1 + (\frac{dy}{dx})^2} \; dx = \sqrt{1 + (\frac{dx}{dy})^2} \; dy$ (use the simpler one)

e.g. The graph of $y = \sqrt{x}$ from $x = 2$ to $x = 6$ is revolved around the x-axis.
      Find the surface area of revolution.

$$dA = 2\pi r \, dL = 2\pi y \cdot \sqrt{1 + (\frac{dy}{dx})^2} \; dx$$

$$= 2\pi y \cdot \sqrt{1 + \frac{1}{4} \cdot \frac{1}{x}} \; dx$$

$$= 2\pi \sqrt{x} \cdot \sqrt{\frac{4x+1}{4x}} \; dx = \pi \cdot \sqrt{4x + 1} \; dx$$

$$\int_2^6 dA = \int_2^6 \pi \cdot \sqrt{4x + 1} \; dx$$

$$= \frac{\pi}{4} \int_2^6 \sqrt{4x + 1} \; d(4x + 1)$$

$$= \frac{\pi}{4} \cdot \frac{(4x+1)^{\frac{3}{2}}}{\frac{3}{2}} \Big|_2^6 = \frac{\pi}{6} \cdot (4x + 1)^{\frac{3}{2}} \Big|_2^6 = \frac{49\,\pi}{3}$$

X. Applications of Integration

5. Work

Work done by a variable force F directed along the x-axis from x = a to x = b is

$$W = \int_a^b F\,dx$$

Hooke's Law

    F = k x   compressing or stretching a spring
                k is a constant ; x is the length change from its natural length.

e.g. Find the work done by compressing or stretching a spring from position a to position b.

$$W = \int_a^b F\,dx = \int_a^b kx\,dx = \frac{kx^2}{2}\Big|_a^b = \frac{k}{2}(b^2 - a^2)$$

e.g. Find the work done by pumping water from containers.

    W = mg($h_2 - h_1$)   m: mass (kg) , mg: weight (N) , mgh: potential energy (J)
                                  g: 9.8m/s^2 , D: density of water (1000 kg/m^3)

    A conical tank with height 10 m and base radius 5 m is filled with water. Find the work required to empty the tank by pumping the water to a level of 4 m above the top of the tank.

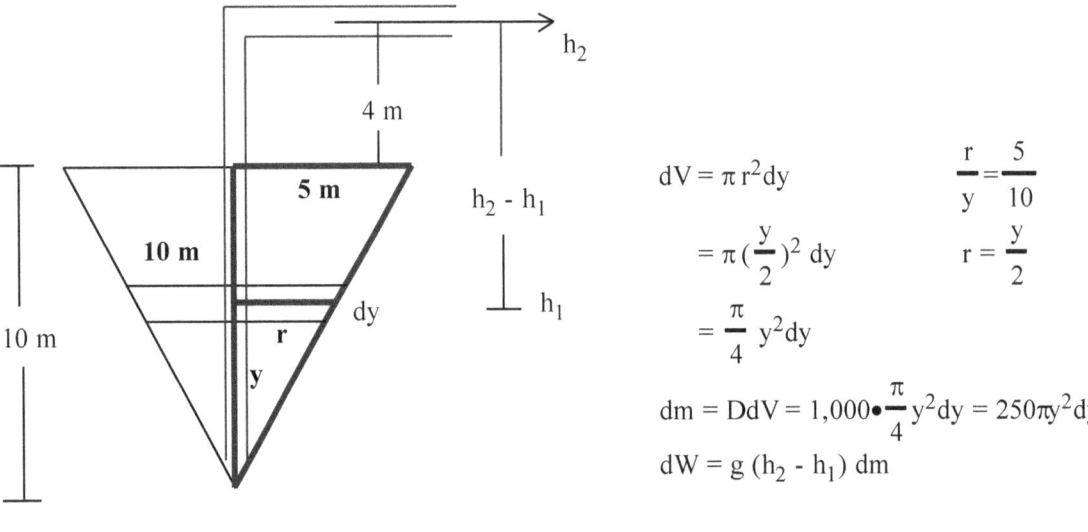

$$dV = \pi r^2 dy \qquad\qquad \frac{r}{y} = \frac{5}{10}$$

$$= \pi\left(\frac{y}{2}\right)^2 dy \qquad r = \frac{y}{2}$$

$$= \frac{\pi}{4} y^2 dy$$

$$dm = D\,dV = 1{,}000\bullet\frac{\pi}{4} y^2 dy = 250\pi y^2 dy$$

$$dW = g\,(h_2 - h_1)\,dm$$

$$W = \int dW = \int g\,((h_2 - h_1)\,dm$$

$$= \int_0^{10} 9.8\,(14 - y)\cdot 250\pi y^2 dy = 2450\pi \int_0^{10}(14 - y)\cdot y^2 dy$$

$$= 2450\pi\left(\frac{14y^3}{3} - \frac{y^4}{4}\right)\Big|_0^{10}$$

$$\approx 1.668 \times 10^7 \text{ J}$$

6. Variable-Density Mass

(1) One-Dimension Mass

$$M = \int_a^b D(x)dx$$ D(x) is the linear density

The density varies along the line.

(2) Two-Dimension Mass

$$M = \int D(x,y)dA$$ D(x, y) is the area density

The density varies by the position in the plane.

(3) Three-Dimension Mass

$$M = \int D(x,y,z)dV$$ D(x, y, z) is the density

The density varies by the position in the space.

e.g. The linear density of a 10-m-long rod is $D(x) = 1 + \frac{x}{10}$ kg/m.
Find the mass of the rod.

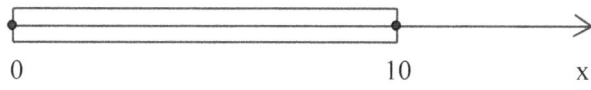

$$M = \int_0^{10} D(x)dx = \int_0^{10}(1 + \frac{x}{10})dx$$

$$= (x + \frac{x^2}{20}) \Big|_0^{10} = 15 \text{ kg}$$

e.g. Variable – Density Population

The population density of a city is $D(r) = \frac{10,000}{1+r^2}$, where r is the distance (miles) from the center of the city. Find the population within 10 miles from the center of the city.

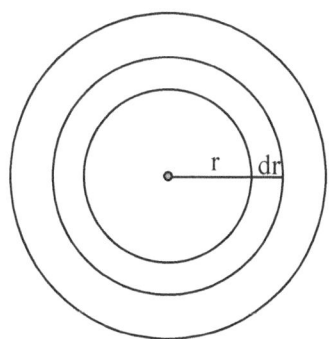

ring sector: $dA = 2\pi\, r\, dr$

$$P = \int D(r)dA = \int_0^{10} \frac{10{,}000}{1+r^2} \cdot 2\pi r\, dr \qquad\qquad dA = 2\pi r\, dr$$

$$= 10{,}000\pi \int_0^{10} \frac{d(1+r^2)}{1+r^2}$$

$$= 10{,}000\pi \cdot \ln(1 + r^2) \Big|_0^{10}$$

$$= 10{,}000\pi \cdot \ln 101 \approx 145{,}000$$

7. The Net Change

The definite integral of the rate of change over an interval is the net change.

If f is a function of t and the rate of change $r(t) = \dfrac{df}{dt}$, then the net change of f(t) as t varies from a to b is

$$\int_a^b r(t)dt = f(b) - f(a)$$

e.g. In physics, the net change in position is called displacement:

$$s(b) - s(a) = \int_a^b v(t)dt$$

e.g. Suppose the current population of a town is 2 million and the rate of growth is $r(t) = 500^{0.2t}$. Find the population of the town in 10 years.

The net change of the population is

$$\int_0^{10} r(t)dt = \int_0^{10} 500^{0.2t}\, dt \approx 201{,}000 \qquad\qquad \text{(use graphing calculator)}$$

The population of the town in 10 years is

$$2{,}000{,}000 + 201{,}000 = 2{,}201{,}000$$

If you know motion, you understand calculus, and vice versa.

In physics, a net change in position is called displacement:

$$s(b) - s(a) = \int_a^b v(t)dt$$

Distance traveled is $\int_a^b |v(t)|dt$, where $|v(t)|$ is called speed.

1. Motion in One-Dimension

A particle moves along a straight line.

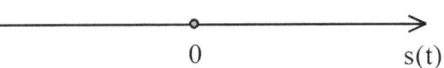

$0 \qquad\qquad s(t)$

Position: $s = s(t)$

Velocity: $v(t) = \dfrac{ds}{dt}$ Speed: $|v(t)|$

Acceleration: $a(t) = \dfrac{dv}{dt} = \dfrac{d^2s}{dt^2}$

(1) $v > 0$, moving to the right.
        $v < 0$, moving to the left.

(2) $a > 0$, v is increasing.
        $a < 0$, v is decreasing.
        Note: $v > 0$ and v is increasing are different concepts.

(3) v and a have the same sign, speed is increasing.
        v and a have opposite signs, speed is decreasing.

(4)   When v becomes 0 and a is not 0, the particle is stopped and changes direction.

Position Curves s (t):

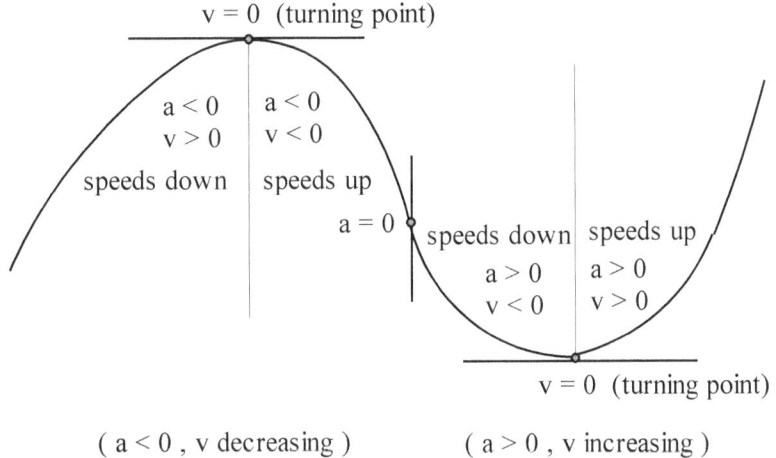

e.g. $s(t) = t^3 - 9t^2 + 15t$
                  compare the graph of $s(t)$ and the graph of $v(t)$.

(a) Solve $s(t) = t^3 - 9t^2 + 15t = 0$, $t = 0,\ t = 2.2,\ t = 6.8$
             $v(t) = s'(t) = 3t^2 - 18t + 15 = 0$ $t = 1,\ t = 5$
             $a(t) = v'(t) = 6t - 18 = 0$ $t = 3$

(b) Position Curve:

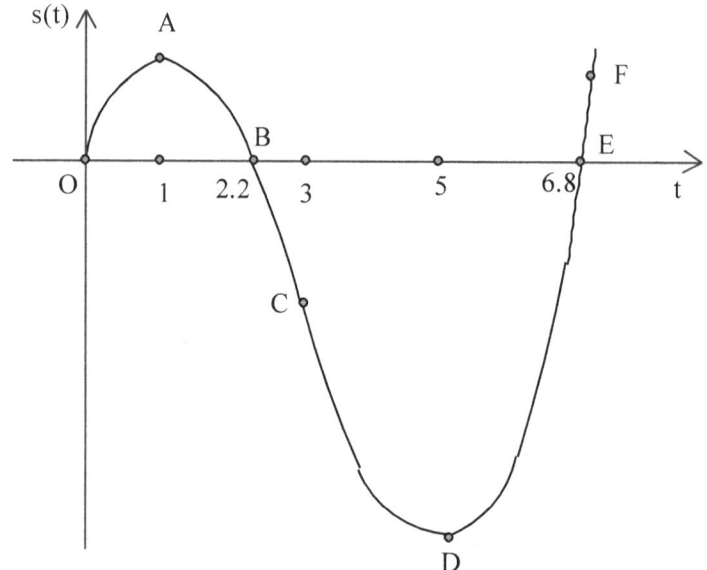

O: the origin at $t = 0$.

$O \to A$: s increasing, $v > 0$, moving to the right ,
            concave down , $a < 0$, v decreasing.

A: stopped, $v = 0$.

$A \to B \to C$: s decreasing, $v < 0$, moving to the left ,
            concave down , $a < 0$, v decreasing.

B: passing the origin at $t = 2.2$.

C: the point of inflection, $a = 0$.

$C \to D$: s decreasing, $v < 0$, moving to the left ,
            concave up , $a > 0$, v increasing.

D: stopped, $v = 0$.

$D \to E \to F$: s increasing, $v > 0$, moving to the right ,
            concave up , $a > 0$, v increasing.

E: passing the origin at $t = 6.8$.

(c)   Velocity Curve:

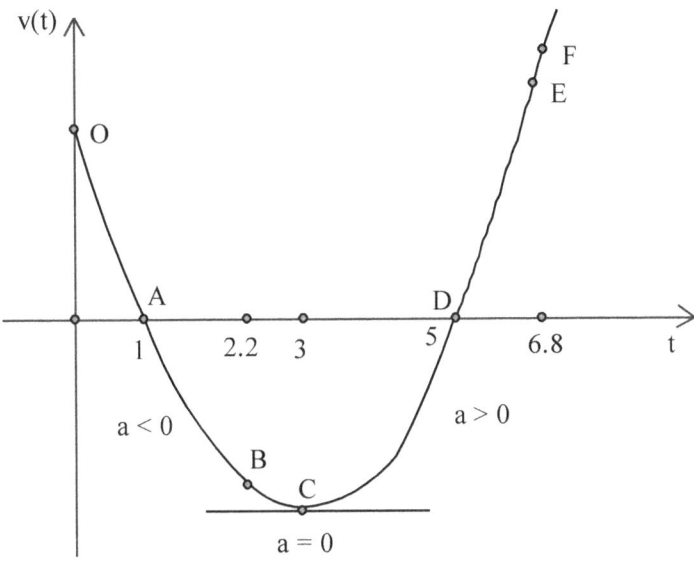

$$v(t) = s'(t) = 3t^2 - 18t + 15$$

O: original velocity $v(0) = 15$.

$O \rightarrow A$: $v > 0$, moving to the right ,
           v decreasing, $a < 0$,   speed decreasing (v and a have opposite signs).

A: $v = 0$, stopped and changing direction.

$A \rightarrow B \rightarrow C$: $v < 0$, moving to the left ,
           v decreasing, $a < 0$,   speed increasing (v and a have the same sign).

C: $a = 0$.   a local minimum for velocity v , but a local maximum for speed $|v(t)|$.

$C \rightarrow D$: $v < 0$, moving to the left ,
           v increasing, $a > 0$,   speed decreasing.

D: $v = 0$, stopped and changing direction.

$D \rightarrow E \rightarrow F$: $v > 0$, moving to the right ,
           v increasing, $a > 0$,   speed increasing.

The region between the curve of $v(t)$ and t-axis over the interval $[t_1 , t_2]$ is the displacement

$$\int_{t_1}^{t_2} v(t)dt$$

Its absolute value $\int_{t_1}^{t_2} |v(t)|dt$ is equal to the area of the regions between the function v(t) and the t-axis. It stands for the total distance traveled.

(d) Use function $v(t)$ to find the time and the position of the particle when it is farthest to the left.

When $v < 0$, the particle is moving to the left . It reaches the farthest at $t = 5$ where $v(5) = 0$, then it turns back to the right .

$$s(5) = s(0) + \int_0^5 v(t)dt$$

$$= 0 + \int_0^5 (3t^2 - 18t + 15)dt \qquad \text{given } s(0) = 0$$

$$= [\, t^3 - 9t^2 + 15t \,]\, _0^5$$

$$= -25$$

(e) Use function $v(t)$ to find the total distance traveled by the particle from $t = 0$ to $t = 5$.

Partition the interval $[0, 5]$ at 1 to separate the positive and negative regions.

$$\int_0^5 |v(t)|dt = \int_0^1 v(t)dt - \int_1^5 v(t)dt \qquad \text{since } \int_1^5 v(t)dt \text{ is negative,}$$
$$\qquad\qquad\qquad\qquad\qquad\qquad\qquad\quad - \int_1^5 v(t)dt \text{ is positive}$$

$$= \int_0^1 (3t^2 - 18t + 15)\, dt - \int_1^5 (3t^2 - 18t + 15)\, dt$$

$$= [\, t^3 - 9t^2 + 15t \,]\, _0^1 - [\, t^3 - 9t^2 + 15t \,]\, _1^5$$

$$= 7 - (-25 - 7) = 39$$

2. Motion in Two-Dimension

A particle moves along a curve in the plane.

(1) Vectors

A vector is a line segment with direction, denoted by $\overrightarrow{AB}$, $\vec{R}$ or $\mathbf{R}$.

Two vectors are equal if they have the same length and direction.

e.g.

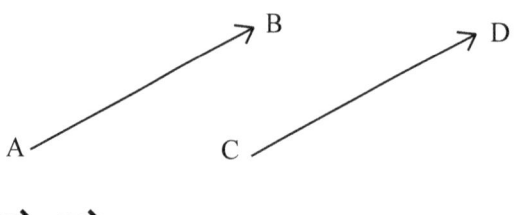

$\overrightarrow{AB} = \overrightarrow{CD}$ implies $AB = CD$ and $\overline{AB} \parallel \overline{CD}$.

A vector can move freely in the plane as long as it is parallel to the original direction.

A vector **R** in the plane can be resolved into two components **Rx** and **Ry**

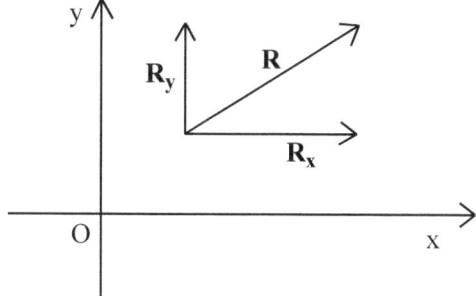

The standard position of a vector is where the initial point is the origin.

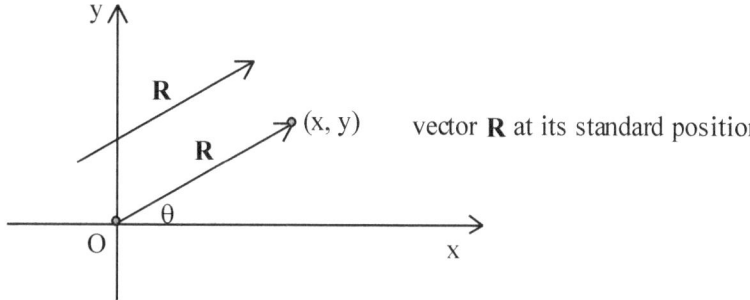

vector **R** at its standard position

The vector **R** at its standard position is called the position vector.

A vector can be expressed by its components:

$$\mathbf{R} = \langle x, y \rangle$$ x is the length of **Rx**, y is the length of **Ry**

The **magnitude** of a vector is the length of the vector segment, denoted by the symbol |**R**|.

$$|\mathbf{R}| = \sqrt{x^2 + y^2}$$

Unit Vector: $\mathbf{u} = \dfrac{\mathbf{R}}{|\mathbf{R}|} = \langle \cos\theta, \ \sin\theta \rangle$ The unit vector is the direction of the vector.

Standard Unit Vectors:

> **i** is the horizontal unit vector, $\mathbf{i} = \langle 1, 0 \rangle$.
> **j** is the vertical unit vector, $\mathbf{j} = \langle 0, 1 \rangle$.

A vector can also be expressed by the standard unit vectors:

$$\mathbf{R} = x\mathbf{i} + y\mathbf{j}$$

$$\mathbf{u} = \cos\theta \ \mathbf{i} + \sin\theta \ \mathbf{j}$$

Addition of Vectors

(1) Geometric Method

$$R = R_1 + R_2$$

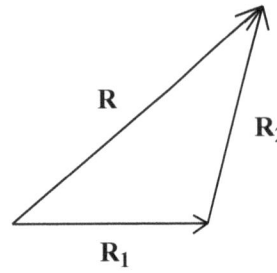 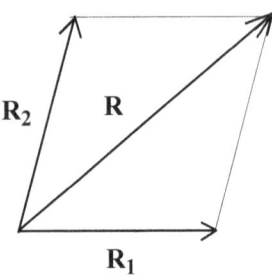

Triangle Law   Parallelogram Law

The sum R is called the resultant vector.

(2) Algebraic Method

$$R_1 = \langle x_1, \ y_1 \rangle, \qquad R_2 = \langle x_2, \ y_2 \rangle$$

$$R = R_1 + R_2 = \langle x_1 + x_2, \ y_1 + y_2 \rangle$$

or $\quad (x_1\mathbf{i} + y_1\mathbf{j}) + (x_2\mathbf{i} + y_2\mathbf{j}) = (x_1 + x_2)\mathbf{i} + (y_1 + y_2)\mathbf{j}$

R and $-R$ are opposite vectors. They have the same length but opposite directions.

Subtraction of a vector is same as the addition of its opposite vector.

Multiplication of a vector by a scalar

(1) Geometric Method

$c\mathbf{R}$ has the same direction as $\mathbf{R}$ and its length is $c|\mathbf{R}|$.

(2) Algebraic Method

$$R = \langle x, \ y \rangle$$

$$cR = \langle cx, \ cy \rangle$$

or $\quad c(x\mathbf{i} + y\mathbf{j}) = cx\mathbf{i} + cy\mathbf{j}$

Dot Product

$$\mathbf{R_1} = \langle x_1,\ y_1 \rangle, \qquad \mathbf{R_2} = \langle x_2,\ y_2 \rangle$$

$$\mathbf{R_1 \cdot R_2} = x_1 \cdot x_2 + y_1 \cdot y_2$$

The dot product of two vectors is a number, not a vector.

Angle between Two Vectors

$$\cos\theta = \frac{\mathbf{R1 \cdot R2}}{|\mathbf{R1}||\mathbf{R2}|}$$

Vectors $\mathbf{R_1}$ and $\mathbf{R_2}$ are perpendicular if and only if $\mathbf{R_1 \cdot R_2} = 0$.

e.g. Find the unit vectors that are tangent and normal to the curve $y = \sin x$ at the point $x = \dfrac{\pi}{3}$.
The slope of the unit vector tangent to the curve is the derivative.

$$m_1 = y' = \cos x \big|_{x = \frac{\pi}{3}} = \frac{1}{2}$$

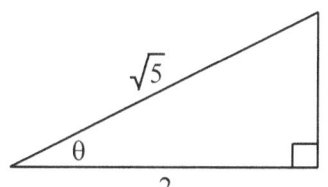

$$|\mathbf{R}| = \sqrt{1^2 + 2^2} = \sqrt{5}$$

The unit vector tangent to the curve at the point $x = \dfrac{\pi}{3}$ is

$$\mathbf{u} = \cos\theta\, \mathbf{i} + \sin\theta\, \mathbf{j} = \frac{2}{\sqrt{5}}\mathbf{i} + \frac{1}{\sqrt{5}}\mathbf{j} \qquad \text{or} \qquad \mathbf{u} = -\frac{2}{\sqrt{5}}\mathbf{i} - \frac{1}{\sqrt{5}}\mathbf{j}$$

The unit vector normal to the curve is perpendicular to the tangent.

$$m_2 = \frac{-1}{m_1} = -2$$

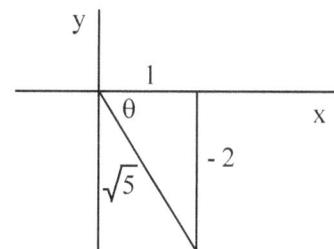

The unit vector normal to the curve at the point $x = \dfrac{\pi}{3}$ is

$$\mathbf{u} = \cos\theta\, \mathbf{i} + \sin\theta\, \mathbf{j} = \frac{1}{\sqrt{5}}\mathbf{i} - \frac{2}{\sqrt{5}}\mathbf{j} \qquad \text{or} \qquad \mathbf{u} = -\frac{1}{\sqrt{5}}\mathbf{i} + \frac{2}{\sqrt{5}}\mathbf{j}$$

(2) Position Vector, Velocity Vectors and Acceleration Vectors

In motion, the position vector $\mathbf{R}(x, y)$ is considered as parametric functions where $x(t)$ and $y(t)$ are functions of time t.

The velocity vector is the derivative of the vector function $\mathbf{R}(x, y)$ — the position vector.

$$\mathbf{v} = \frac{d\mathbf{R}}{dt} = \frac{dx}{dt}\mathbf{i} + \frac{dy}{dt}\mathbf{j}$$

or $\quad \mathbf{v}(t) = \langle \frac{dx}{dt}, \frac{dy}{dt} \rangle = \langle v_x, v_y \rangle$

The direction of the $\mathbf{v}$ is the tangent of the curve at the point:

$$\frac{dy}{dx} = \frac{\frac{dy}{dt}}{\frac{dx}{dt}} = \frac{y'(t)}{x'(t)}$$

The magnitude of the vector $\mathbf{v}$ is the speed

$$|\mathbf{v}| = \sqrt{(\frac{dx}{dt})^2 + (\frac{dy}{dt})^2} = \sqrt{(v_x)^2 + (v_y)^2}$$

The direction of motion can also be expressed as $\dfrac{\mathbf{v}}{|\mathbf{v}|}$, which is a unit vector.

The total distance traveled

$$\int_{t_1}^{t_2} |\mathbf{v}|dt = \int_{t_1}^{t_2} \sqrt{(\frac{dx}{dt})^2 + (\frac{dy}{dt})^2}\, dt$$

The displacement — the net change of positions

$$x(t_2) - x(t_1) = \int_{t_1}^{t_2} v_x dt \qquad \text{or} \qquad x(t_2) = x(t_1) + \int_{t_1}^{t_2} v_x dt$$

$$y(t_2) - y(t_1) = \int_{t_1}^{t_2} v_y dt \qquad \text{or} \qquad y(t_2) = y(t_1) + \int_{t_1}^{t_2} v_y dt$$

The acceleration vector is the derivative of the velocity vector $\mathbf{v}(x, y)$.

$$\mathbf{a} = \frac{d\mathbf{v}}{dt} = \frac{d^2x}{dt^2}\mathbf{i} + \frac{d^2y}{dt^2}\mathbf{j}$$

or $\quad \mathbf{a}(t) = \langle \frac{d^2x}{dt^2}, \frac{d^2y}{dt^2} \rangle = \langle a_x, a_y \rangle$

e.g. The velocity of a particle moving in the plane has components:

$$\frac{dx}{dt} = t^2 - 3t + 2 \ \text{ and } \ \frac{dy}{dt} = 2t - 1$$

At time $t = 0$, the position of the particle is $(2, -5)$.

(a) Find all the values of t at which the tangent to the path of the particle is vertical.

The tangent is vertical when $\frac{dy}{dx}$ is undefined.

$$x'(t) = \frac{dx}{dt} = 0 \ \text{ and } \ y'(t) = \frac{dy}{dt} \neq 0$$

$x'(t) = t^2 - 3t + 2 = 0 \qquad\qquad t = 1 \text{ and } t = 2$

$y'(t) = 2t - 1, \qquad\qquad y'(1) = 1 \neq 0 \text{ and } y'(2) = 3 \neq 0$

Therefore the tangent is vertical when $t = 1$ and $t = 2$.

(b) Write the equation for the tangent line to the path of the particle at $t = 3$.

$$\text{slope } m = \frac{dy}{dx}\Big|_{t=3} = \frac{\frac{dy}{dt}}{\frac{dx}{dt}} = \frac{2t-1}{t^2-3t+2} = \frac{5}{2}$$

$$x(3) = x(0) + \int_0^3 x'(t)dt = 2 + \int_0^3 (t^2 - 3t + 2)dt$$

$$= 2 + \left[\frac{t^3}{3} - \frac{3t^2}{2} + 2t\right]\Big|_0^3 = 2 + \frac{3}{2} = 3.5$$

$$y(3) = y(0) + \int_0^3 y'(t)dt = -5 + \int_0^3 (2t - 1)dt$$

$$= -5 + \left[\frac{2t^2}{2} - t\right]\Big|_0^3 = -5 + 6 = 1$$

The equation of the tangent line:

$$y - 1 = \frac{5}{2}(x - 3.5)$$

(c) Find the speed of the particle at $t = 3$.

$$|\mathbf{v}| = \sqrt{(x'(t))^2 + (y'(t))^2}$$

$$= \sqrt{(x'(3))^2 + (y'(3))^2}$$

$$= \sqrt{(2)^2 + (5)^2}$$

$$= \sqrt{29}$$

e.g. Projectile Motion

The position vector is $\mathbf{R}(t) = x\mathbf{i} + y\mathbf{j}$, where

$$x = 2 - 2\cos t, \quad y = 4 - 4\cos 2t \qquad\qquad 0 \le t \le \pi$$

(a) Write the equation of the curve in x and y, and sketch the curve.

$$
\begin{array}{ll}
y = 4 - 4\cos 2t & x = 2 - 2\cos t \\
\quad = 4 - 4(2\cos^2 t - 1) & (x - 2)^2 = 4\cos^2 t \\
\quad = 8 - 8\cos^2 t & 2(x - 2)^2 = 8\cos^2 t
\end{array}
$$

$$y = -2(x - 2)^2 + 8 \quad \text{or} \quad y = -2x^2 + 8x \qquad \text{where } x \ge 0 , \ y \ge 0$$

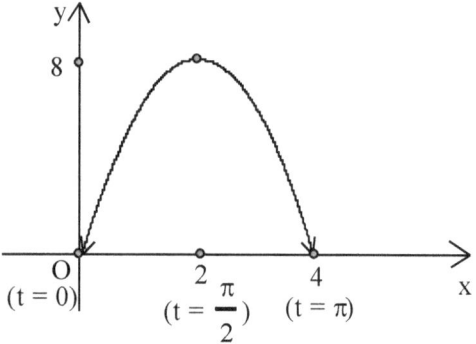

(b) Find the velocity and acceleration vectors.

$$\frac{dx}{dt} = 2\sin t , \qquad \frac{dy}{dt} = 8\sin 2t$$

$$\mathbf{v}(t) = \langle \frac{dx}{dt} , \frac{dy}{dt} \rangle = \langle 2\sin t , \ 8\sin 2t \rangle$$

$$\frac{d^2 x}{dt^2} = 2\cos t , \qquad \frac{d^2 y}{dt^2} = 16\cos 2t$$

$$\mathbf{a}(t) = \langle \frac{d^2 x}{dt^2} , \frac{d^2 y}{dt^2} \rangle = \langle 2\cos t, 16\cos 2t \rangle$$

(c) Find the maximum height of the projectile.

Method 1:   solve $\dfrac{dy}{dx} = 0$

$$\frac{d}{dx}(-2x^2 + 8x) = -4x + 8 = 0$$

$$x = 2$$

$$y = -2x^2 + 8x$$
$$= -2(2)^2 + 8(2)$$
$$= 8 \quad \text{(maximum height)}$$

Method 2:   solve $\frac{dy}{dt} = 0$

$$\frac{d}{dt}(4 - 4\cos2t) = 8\sin2t = 0$$

$$\sin2t = 0, \quad t = \frac{\pi}{2}$$

$$y(\tfrac{\pi}{2}) = 4 - 4\cos2(\tfrac{\pi}{2}) = 8$$

(d) Find when the projectile has the maximum speed and minimum speed.

$$|\mathbf{v}| = \sqrt{(\frac{dx}{dt})^2 + (\frac{dy}{dt})^2}$$
$$= \sqrt{(2\sin t)^2 + (8\sin2t)^2}$$
$$= \sqrt{4\sin^2 t + 64\sin^2 2t}$$

Use a graphing calculator to find the maximum and minmum values of the function

$$f(t) = 4\sin^2 t + 64\sin^2 2t$$

Maximum speed at t = 0.793 and t = 2.348 .

Minimum speed at t = 1.571 = $\frac{\pi}{2}$.

Tip: The window dimensions: $x_{min} = 0$, $x_{max} = \pi$, $y_{min} = 0$, $y_{max} = 100$

(e) Find the total distance traveled by the projectile from $t = 0$ to $= \pi$.

Method 1: $\int_0^\pi |v(t)|dt$
$$= \int_0^\pi \sqrt{4\sin^2 t + 64\sin^2 2t} \; dt \approx 16.819$$

Method 2: Find the arc length of $y = -2x^2 + 8x$ from x = 0 to x = 4.

$$\int_0^4 \sqrt{1 + (\frac{dy}{dx})^2} \; dx = \int_0^4 \sqrt{1 + (-4x + 8)^2}dx$$
$$= \int_0^4 \sqrt{16x^2 - 64x + 65} \; dx$$

$$\approx 16.819$$

X. Mean Value Theorems

1. The Intermediate Value Theorem

If f(x) is continuous on the closed interval [a, b], then every value between f(a) and f(b) exists and corresponds to some c in (a, b) .

If f(a) < f(b) , then there is some c in (a, b) such that f(a) < f(c) < f(b).   or

If f(a) > f(b) , then there is some c in (a, b) such that f(a) > f(c) > f(b).

The Intermediate Value Theorem is used to find zeros.
If f(x) is continuous, then any interval on which f(x) changes sign contains at least one root of the equation f(x) = 0.

e.g. $x^2 + 8x + 6 = 0$

$$f(x) = x^2 + 8x + 6$$

$$f(-1) = -1 \text{ and } f(0) = 6$$

Therefore there is at least one root in the interval (-1, 0) .

$$x \approx -0.838$$

This theorem is also used in the graphing calculator, Left Bound? Right Bound?, to find zeros.

e.g.   Show there is at least one root for $x^2 = \sqrt{x+1}$.

Let $f(x) = x^2 - \sqrt{x+1}$

f(x) is continuous.

f(0) = -1 and f(3) = 7, therefore there is at least one root in the interval (0, 3) for $x^2 = \sqrt{x+1}$.

2. The Mean Value Theorem

If f(x) is continuous on the closed interval [a, b] and is differentiable everywhere on the open interval (a, b), then there is at least one number c between a and b such that

$$f'(c) = \frac{f(b)-f(a)}{b-a}$$

$\frac{f(b)-f(a)}{b-a}$ is the average rate of change for function f over the interval [a, b].

$f'(c)$ is the instantaneous rate of change at point c.

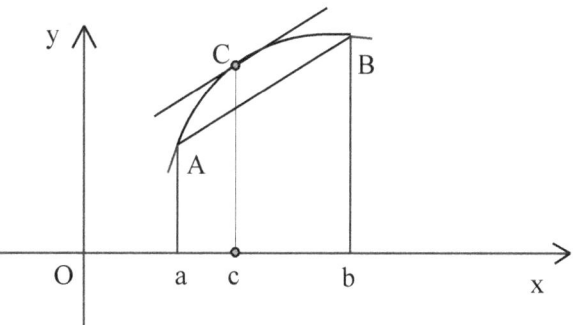

The tangent at point C is parallel to the chord $\overline{AB}$.

In motion problems, if the position $s = f(t)$, then $f(b) - f(a)$ represents the net change in position, displacement, over a time period $(b - a)$. $\frac{f(b)-f(a)}{b-a}$ represents the average velocity. $f'(c)$ is the instantaneous velocity at point c.

e.g. What is the average rate of change of the function $f(x) = x^3 - 2x$ on the closed interval $[0, 2]$ and at which point between 0 and 2 does the function have the same rate of change?

The average rate of change: $\frac{f(2)-f(0)}{2-0} = \frac{4}{2} = 2$

The instantaneous rate of change: $f'(x) = 3x^2 - 2$

Solve: $f'(x) = 3x^2 - 2 = 2$

$x^2 = \frac{4}{3}$

$x = \frac{2\sqrt{3}}{3} \approx 1.155$ since x is between 0 and 2.

Rolle's Theorem

Suppose that $f(x)$ is continuous on the closed interval [a, b] and is differentiable everywhere on the open interval (a, b). If

$$f(a) = f(b) = 0$$

then there is at least one number c in (a, b) at which $f'(c) = 0$.

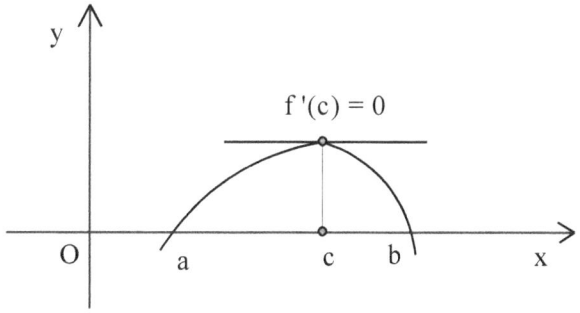

X. Mean Value Theorems

e.g. Show that the equation $x^3 + 2x - 1 = 0$ has exactly one real zero.

$f(x) = x^3 + 2x - 1$ is differentiable at every value of x, and its derivative $f'(x) = 3x^2 + 2$ is never zero. If f had two or more zeros, f' would have a zero between them. Therefore f has at most one zero. On the other hand, $f(-1) = -4$ is negative, $f(1) = 2$ is positive and f is continuous, therefore f has at least one zero between -1 and 1. We have proved that f has exactly one zero.

3. Average Value (The Mean Value Theorem for Integrals)

If f(x) is continuous on the closed interval [a, b], then at some point c in the interval [a, b] such that

$$f(c) = \frac{1}{b-a} \int_a^b f(x)dx$$

The number $f(c)$ is the average value of $f(x)$ on the interval [a, b] .

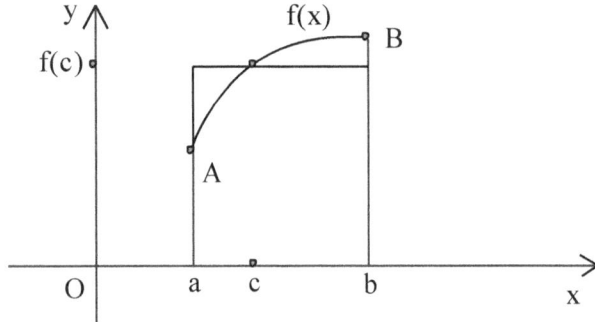

For positive function f , the area between the graph of f and x-axis on [a, b] is equal to the area of the rectangle formed by the height f(c) and the base (b - a).

In motion problems, if the velocity $v = f(t)$, then $\int_a^b f(t)dt$ represents the total displacement over a time period $(b - a)$. The average velocity is the total displacement divided by the time $(b - a)$.

$$\bar{v} = \frac{1}{b-a} \int_a^b f(t)dt$$

The Theorem tells us there is at least one moment c in the period [a, b] such that the instantaneous velocity $f(c)$ equals the average velocity $\bar{v}$.

e.g. Find the average value of $f(x) = x^2$ from $x = 0$ to $x = 3$.

$$\frac{1}{3-0} \int_0^3 x^2 \, dx$$

$$= \frac{1}{3} \left(\frac{x^3}{3}\right) \Big|_0^3 = 3$$

(Note: The average value of $f(x)$ is not equal to $\dfrac{f(0)+f(3)}{2} = \dfrac{0+9}{2} = 4.5$)

A differential equation is an equation that contains derivatives. To solve a differential equation is to determine its original function.

1. Separation of Variables

A separable equation can be written in the form

$$g(y)dy = f(x)dx \qquad \text{or} \qquad f(x)dx + g(y)dy = 0$$

e.g   Find the general solution.

$$\frac{dy}{dx} = \frac{3x^2 + 2x}{3y^2}$$

Separate the variables:

$$3y^2 dy = (3x^2 + 2x)dx$$

Integrate both sides:

$$\int 3y^2 dy = \int (3x^2 + 2x)dx$$

$$\frac{3y^3}{3} = \frac{3x^3}{3} + \frac{2x^2}{2} + C$$

$$y^3 = x^3 + x^2 + C$$

$$y = \sqrt[3]{x^3 + x^2 + C}$$

e.g.   Solve $\dfrac{dy}{dx} = \dfrac{e^{2x-y}}{e^{x+y}}$.

$$\frac{dy}{dx} = \frac{e^{2x-y}}{e^{x+y}} = \frac{e^x}{e^{2y}}$$

$$e^{2y} dy = e^x dx$$

$$\int e^{2y} dy = \int e^x dx$$

$$\frac{1}{2} e^{2y} = e^x + C_1$$

$$e^{2y} = 2e^x + C$$

e.g.   Solve $\dfrac{dy}{dx} = 2xy$ when the initial condition is $y(0) = 5$.

$$\frac{dy}{y} = 2xdx$$

$$\int \frac{dy}{y} = \int 2xdx$$

$$\ln y = \frac{2x^2}{2} + C_1 = x^2 + C_1$$

$$y = e^{x^2 + C_1} = e^{x^2} \cdot e^{C_1} = C e^{x^2} \qquad\qquad \text{let } C = e^{C_1}$$

$$y(0) = C e^0 = C = 5$$

The solution is $y = 5e^{x^2}$

2. Homogeneous Equations

A differential equation can be written in the form

$$\frac{dy}{dx} = f\left(\frac{y}{x}\right)$$

Let $v = \dfrac{y}{x}$, then $y = vx$, $\dfrac{dy}{dx} = \dfrac{d(vx)}{dx} = x\dfrac{dv}{dx} + v$

The equation becomes a separable equation.

e.g $$\frac{dy}{dx} = \frac{x^2 - y^2}{xy}$$

Write the equation in the homogeneous form:

$$\frac{dy}{dx} = \frac{1 - \left(\frac{y}{x}\right)^2}{\left(\frac{y}{x}\right)}$$

$$\frac{d(vx)}{dx} = \frac{1 - v^2}{v} \qquad\qquad \text{let } v = \frac{y}{x}$$

$$x\frac{dv}{dx} + v = \frac{1 - v^2}{v}$$

$$x\frac{dv}{dx} = \frac{1 - v^2}{v} - v = \frac{1 - v^2 - v^2}{v} = \frac{1 - 2v^2}{v}$$

$$\frac{vdv}{1-2v^2} = \frac{dx}{x}$$

$$-\frac{1}{4} \cdot \frac{d(1-2v^2)}{1-2v^2} = \frac{dx}{x}$$

$$\int \frac{dx}{x} + \frac{1}{4} \int \frac{d(1-2v^2)}{1-2v^2} = 0$$

$$\ln x + \frac{1}{4}\ln(1-2v^2) = C_1, \qquad 4\ln x + \ln(1-2v^2) = C_2$$

$$\ln[x^4 \cdot (1-2v^2)] = C_2, \qquad x^4 \cdot (1-2v^2) = e^{C_2} = C$$

$$x^4 \cdot \left(1 - 2\left(\frac{y}{x}\right)^2\right) = C$$

$$x^4 - 2x^2y^2 = C \qquad\qquad \text{the general solution}$$

3. Linear First-Order Equations

A differential equation can be written in the form

$$\frac{dy}{dx} + P(x)y = Q(x)$$

Let $v(x) = e^{\int P(x)dx}$ (Any antiderivative of P(x) will do. Omit the constant C)

$$y = \frac{1}{v(x)} \int v(x) \cdot Q(x)dx \qquad (\text{Remember it by two parts } \frac{1}{v(x)} \int v(x) \text{ and } Q(x)dx.)$$

e.g   Solve $x\frac{dy}{dx} + 2y = 2x^2$ when the initial condition is $y(0) = 5$.

Write the equation in the standard form:

$$\frac{dy}{dx} + \frac{2}{x}y = 2x$$

Let $v(x) = e^{\int \frac{2}{x}dx} = e^{2\ln x} = x^2$ Omit the constant C

$$y = \frac{1}{v(x)} \int v(x) \cdot Q(x)dx$$

$$= \frac{1}{x^2} \int x^2 \cdot 2xdx$$

XI. Differential Equations

$$= \frac{1}{x^2} \cdot \frac{2x^4}{4} + C$$

$$= \frac{x^2}{2} + C$$

$$y(0) = \frac{0^2}{2} + C = 5, \quad C = 5$$

The solution is $y = \frac{x^2}{2} + 5$.

4. Exponential Growth and Decay

When the rate of change is proportional to the variable itself, it leads to the differential equation in the form

$$\frac{dy}{dt} = ky$$

where k is a constant.

If $k > 0$, it is called exponential growth; if $k < 0$, it is called exponential decay.

Half-life: $\qquad y = \frac{1}{2} y_0$

Solve the equation by separation of variables.

$$\int \frac{dy}{y} = \int k\, dt$$

$$\ln y = kt + C_1$$

$$y = e^{kt + C_1} = e^{kt} \cdot e^{C_1} = C e^{kt} \qquad \text{let } C = e^{C_1}$$

$$y(0) = C e^0 = C = y_0$$

$$y = y_0 e^{kt} \qquad\qquad y_0 \text{ is the initial value when } t = 0.$$

Interest Compounding

Interest compounded n times a year

$$A = A_0 \left(1 + \frac{r}{n}\right)^{nt} \qquad t : \text{number of years}$$

Interest compounded continuously, $n \to \infty$

$$\lim_{n\to\infty} A_0 \left(1 + \frac{r}{n}\right)^{nt} = A_0 \lim_{n\to\infty} \left(1 + \frac{r}{n}\right)^{nt} = A_0 e^{rt}$$

e.g.   Compare the money compounded annualy, monthly, and continuously.

Original investment $A_0 = \$1000$
Annual interest rate $r = 5\%$
Number of years $t = 8$

(a) Interest compounded annually

$$A = A_0 (1 + r)^t$$
$$= 1000 (1 + 0.05)^8$$
$$= 1477.46$$

(b) Interest compounded monthly

$$A = A_0 (1 + \frac{r}{12})^{12t}$$
$$= 1000 (1 + \frac{0.05}{12})^{12 \cdot 8}$$
$$= 1490.59$$

(c) Interest compounded continuously

$$A = A_0 e^{rt}$$
$$= 1000 e^{0.05 \cdot 8}$$
$$= 1491.82$$

Newton's Law of Cooling

The rate of cooling of an object is proportion to the temperature difference between the object and its surroundings.

$$\frac{dT}{dt} = -k(T - T_S)$$

T is the temperature of the object, T_S is the temperature of the surroundings, and k is a constant. Since $\frac{dT}{dt}$ is negative, we use $-k$.

Solve the equation by separation of variables.

$$\int \frac{dT}{(T - T_S)} = -\int k\,dt$$

$$\ln(T - T_S) = -kt + C_1$$

$$T - T_S = e^{-kt + C_1} = e^{-kt} \cdot e^{C_1} = C e^{-kt} \qquad \text{let } C = e^{C_1}$$

$$T = C e^{-kt} + T_S$$

$$T_0 = T(0) = C\,e^0 + T_S$$

$$C = T_0 - T_S$$

$$T = (T_0 - T_S)e^{-kt} + T_S$$

or $\qquad \dfrac{T - T_S}{T_0 - T_S} = e^{-kt}$

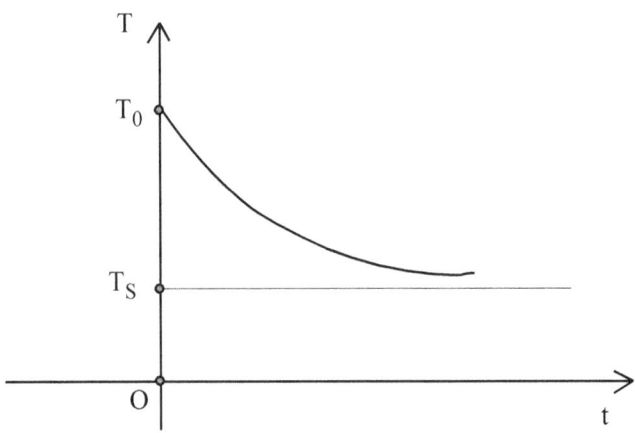

e.g. It takes 20 minutes to cool a cup of coffee from 90°C to 60°C at room temperature of 20°C. How long does it take to cool it from 90°C to 30°C ?

General Solution:

$$\dfrac{T_1 - T_S}{T_0 - T_S} = e^{-kt_1} \qquad\qquad (1)$$

$$\ln\dfrac{T_1 - T_S}{T_0 - T_S} = -kt_1 \qquad\qquad (2)$$

same for T_2 and t_2 : $\quad \ln\dfrac{T_2 - T_S}{T_0 - T_S} = -kt_2 \qquad (3)$

Eq.(3) ÷ Eq.(2) : $\qquad \dfrac{t_2}{t_1} = \dfrac{\ln\dfrac{T_2 - T_S}{T_0 - T_S}}{\ln\dfrac{T_1 - T_S}{T_0 - T_S}}$

Solution to this problem: $T_0 = 90,\ T_S = 20\,,\ T_1 = 60,\ T_2 = 30,\ t_1 = 20$

$$t_2 = 20 \cdot \dfrac{\ln\dfrac{10}{70}}{\ln\dfrac{40}{70}} \approx 70 \text{ min.}$$

The Logistic Model

This model is used for population growth.

$$\frac{dP}{dt} = kP(1 - \frac{P}{P_M})$$

Solution: $P(t) = \frac{P_M}{1+Ce^{-kt}}$ where $C = \frac{P_M - P_0}{P_0}$

where P is the population, P_M is the carrying capacity, which is the maximum population that the environment is capable of sustaining in the long run, and k is a constant.

If P is small compared with P_M , then $\frac{P}{P_M}$ is close to 0 and so $\frac{dP}{dt} = kP$ which is the exponential growth.

e.g   Solve $\frac{dP}{dt} = 0.1P(1 - \frac{P}{1000})$ and $P(0) = 100$

$$\frac{dP}{dt} = \frac{0.1}{1000} P(1000 - P)$$

$$\frac{dP}{P(1000-P)} = \frac{0.1}{1000} dt$$

Integrate by Partial Fractions:

$$\frac{1}{1000} [\int \frac{dP}{P} + \int \frac{dP}{(1000-P)}] = \frac{0.1}{1000} \int dt$$

$$\int \frac{dP}{P} + \int \frac{dP}{(1000-P)} = 0.1 \int dt$$

$$\ln P - \ln(1000 - P) = 0.1t + C_1$$

$$-\ln P + \ln(1000 - P) = -0.1t + C_2$$

$$\frac{1000-P}{P} = e^{-0.1t+C_2}$$

$$\frac{1000}{P} - 1 = Ce^{-0.1t}$$

$$P = \frac{1000}{1+Ce^{-0.1t}}$$

$$(P(0) = \frac{1000}{1+Ce^0} = 100, \ C = 9)$$

$$P = \frac{1000}{1+9e^{-0.1t}}$$

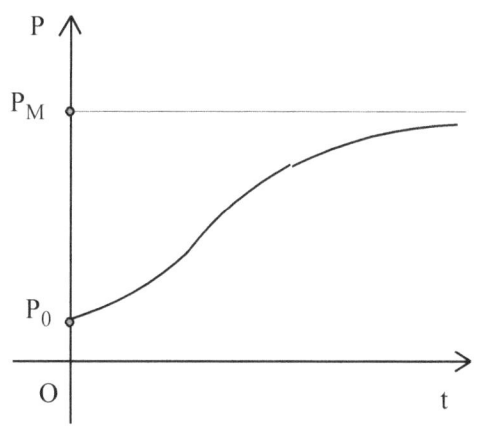

5. Euler's Method (Linear Approximation)

A differential equation is in the form

$$\frac{dy}{dx} = f(x, y)$$

and the initial condition $y(x_0) = y_0$.

Euler's Method gives out an approximation to the value of y in the local interval about x_0 .

$$y_1 = y_0 + \Delta\, y_0 \approx y_0 + f(x_0, y_0\,)dx$$
$$y_2 = y_1 + \Delta\, y_1 \approx y_1 + f(x_1, y_1\,)dx$$
$$y_3 = y_2 + \Delta\, y_2 \approx y_2 + f(x_2, y_2\,)dx$$
$$\cdots\cdots$$

     e.g.   Find the first three approximations for the initial value problem

$$y' = 2x + y \qquad \text{with} \ \ y(1) = 2 \ \text{ and } \ dx = 0.1$$

$$x_0 = 1, \ \ y_0 = 2, \ \ f\big(x_i, \ y_i\big) = 2x_i + y_i$$

$$y_1 = y_0 + (2x_0 + y_0)dx = 2 + (2 \cdot 1 + 2)(0.1) = 2.4 \qquad\qquad x_1 = 1.1$$
$$y_2 = y_1 + (2x_1 + y_1)dx = 2.4 + (2 \cdot 1.1 + 2.4)(0.1) = 2.86 \qquad\quad x_2 = 1.2$$
$$y_3 = y_2 + (2x_2 + y_2)dx = 2.86 + (2 \cdot 1.2 + 2.86)(0.1) = 3.386 \quad\ x_3 = 1.3$$

6. Slope Field

If a differential equation is in the form

$$\frac{dy}{dx} = f(x, y)$$

then every point (x, y) in the plane corresponds to a slope. If we mark each point with a small tangent segment, then these slope indicators form a slope field.

The slope field can help us to visualize the general shape of the solution curves.

There are two types of the questions.

(1) Based on the differential equation and the original condition, sketch the curve of the particular solution.

e.g. Sketch the solution curve for

$$\frac{dy}{dx} = 2x, \qquad y(0) = 1$$

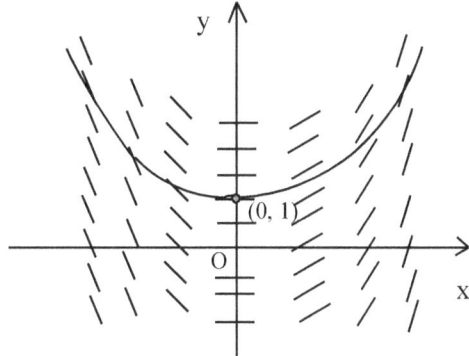

(2) Match the slope fields to the differential equations.

$\frac{dy}{dx} = f(x)$ slope indicators are parallel along any vertical line as in the example above.

$\frac{dy}{dx} = f(y)$ slope indicators are parallel along any horizontal line.

$\frac{dy}{dx} = f\left(\frac{y}{x}\right)$ or $\frac{dy}{dx} = f\left(\frac{x}{y}\right)$ slope indicators are parallel along any line passing through the origin. See the following example.

e.g. $\frac{dy}{dx} = -\frac{x}{2y}$

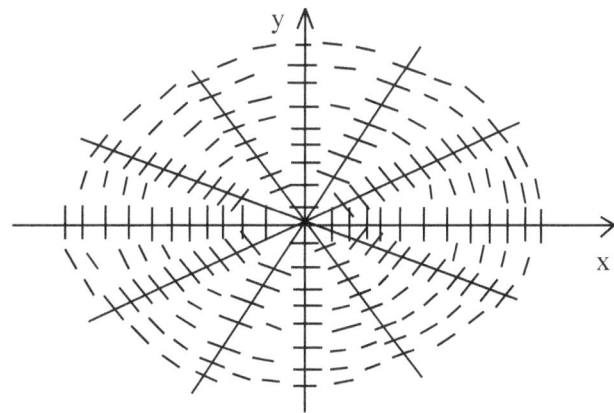

Checking the slopes along the y-axis (x = 0), along the x-axis (y = 0), and along the line y = x , also helps us to identify the equation.

As in the example above, we can also see:
when x = 0, all slopes are equal to zero along the y-axis.
when y = 0, all slopes are undefined along the x-axis.

Part I. Series with Constant Terms

Convergent or not convergent, that is the question.

A **sequence** is a special type of function:

$$a_n = f(n) \qquad \text{where } \{n: \ 1, 2, 3, 4, 5, \cdots \}$$

An infinite sequence converges if it has a limit.

$$\lim_{n \to \infty} a_n = L \qquad L \text{ is a finite number}$$

Otherwise the infinite sequence diverges.

Theorem: If $\lim_{x \to \infty} f(x) = L$, then $\lim_{n \to \infty} f(n) = L$.

$$\text{e.g. } \lim_{x \to \infty} \left(1 + \frac{1}{x}\right)^x = e, \text{ therefore } \lim_{n \to \infty} \left(1 + \frac{1}{n}\right)^n = e$$

Remember that a sequence is a function. Any rule we use to determine the limit of a function is true for the limit of a sequence.

$$\text{e.g. } \lim_{n \to \infty}[f(n) + g(n)] = \lim_{n \to \infty} f(n) + \lim_{n \to \infty} g(n)$$

$$\text{e.g. } \lim_{n \to \infty} \frac{\ln n}{n} \qquad\qquad \frac{\infty}{\infty} \ \ \text{L' Hôpital's Rule}$$

$$= \lim_{n \to \infty} \frac{\frac{1}{n}}{1} = 0$$

Some important limits: (x remains fixed as $n \to \infty$)

$$\lim_{n \to \infty} \frac{\ln n}{n} = 0$$

$$\lim_{n \to \infty} \sqrt[n]{n} = 1$$

$$\lim_{n \to \infty} \sqrt[n]{x} = 1 \qquad\qquad x > 0$$

$$\lim_{n \to \infty} x^n = 0 \qquad\qquad |x| < 1$$

$$\lim_{n \to \infty} \frac{x^n}{n!} = 0 \qquad\qquad \text{any } x$$

$$\lim_{n \to \infty} \left(1 + \frac{x}{n}\right)^n = e^x \qquad\qquad \text{any } x$$

The order of infinities:

$$\ln n < n^a \ (0 < a < 1) < n < n^a \ (a > 1) < a^n \ (a > 1) < n! < n^n$$

A **series** is the sum of a sequence.

$$S_n = \sum_{i=1}^{n} a_i$$

S_n represents the nth partial sum, the sum of the first n terms of a sequence.

An infinite series converges if it has a limit.

$$\lim_{n \to \infty} S_n = S \qquad S \text{ is a finite number}$$

Otherwise the infinite series diverges.

e.g. $\qquad \lim_{n \to \infty} \sum_{i=1}^{n} \frac{1}{2^i} = \frac{1}{2} + \frac{1}{4} + \frac{1}{8} + \frac{1}{16} \cdots = 1 \qquad$ converges

$$\lim_{n \to \infty} \sum_{i=1}^{n} \frac{1}{i} = 1 + \frac{1}{2} + \frac{1}{3} + \frac{1}{4} \cdots \qquad \text{diverges}$$

We are going to learn a number of tests to determine the covergency of an infinite series.

The question always arises: which test should we do first?

The following tests are listed in a preferred order.

1. The nth Term Test

If $\lim_{n \to \infty} a_n \neq 0$, then $\sum_{n=1}^{\infty} a_n$ diverges.

Note: $\lim_{n \to \infty} a_n = 0$ is a necessary condition for the convergency of an infinite series, but not a sufficient condition.

e.g. $\qquad \lim_{x \to \infty} \frac{n}{n+1} = 1$, therefore $\sum_{n=1}^{\infty} \frac{n}{n+1}$ diverges.

e.g. $\qquad \lim_{x \to \infty} \frac{1}{n} = 0$, we can not determine its convergency based on this test.

2. Geometric Series Test

A geometric series with a common ratio r

$$\sum_{n=1}^{\infty} ar^{n-1} = a + ar + ar^2 + \cdots + ar^{n-1} + \cdots$$

if $|r| < 1$, the series converges;   if $|r| \geq 1$, the series diverges.

The nth partial sum $\; S_n = a + ar + ar^2 + \cdots + ar^{n-1} = \dfrac{a(1-r^n)}{1-r} \qquad r \neq 1$

$$\lim_{n \to \infty} S_n = \frac{a}{1-r} \qquad |r| < 1$$

e.g.   Find the limit of $0.252525 \cdots$

$0.252525 \cdots = 0.25 + 0.0025 + 0.000025 \cdots$

$a = 0.25 , \quad r = \dfrac{1}{100} = 0.01 < 1$

$\lim\limits_{n \to \infty} \sum_{n=1}^{\infty} ar^{n-1} = \dfrac{a}{1-r} = \dfrac{0.25}{1-0.01} = \dfrac{0.25}{0.99} = \dfrac{25}{99}$

e.g. $\sum_{n=1}^{\infty} \dfrac{3}{(-2)^n}$

$a = -\dfrac{3}{2}, \quad r = -\dfrac{1}{2} , \quad |r| < 1$

$\lim\limits_{n \to \infty} S_n = \dfrac{a}{1-r} = \dfrac{-\frac{3}{2}}{1-(-\frac{1}{2})} = \dfrac{-\frac{3}{2}}{\frac{3}{2}} = -1$

3. P - Series Test

$\sum_{n=1}^{\infty} \dfrac{1}{n^p} = 1 + \dfrac{1}{2^p} + \dfrac{1}{3^p} + \cdots + \dfrac{1}{n^p} + \cdots$ if $p > 1$, converges

if $p \leq 1$, diverges

Note: In Geometric Series, n is the exponent.
       In P-Series, n is the base.

e.g.   Harmonic Series:

$\sum_{n=1}^{\infty} \dfrac{1}{n} = 1 + \dfrac{1}{2} + \dfrac{1}{3} + \cdots + \dfrac{1}{n} + \cdots$ diverges, since p = 1

e.g.   Telescoping Series:

$\sum_{n=1}^{\infty} \dfrac{1}{n(n+1)} = \lim\limits_{k \to \infty} \sum_{n=1}^{k} \dfrac{1}{n(n+1)}$

$= \lim\limits_{k \to \infty} \sum_{n=1}^{k} \left(\dfrac{1}{n} - \dfrac{1}{n+1} \right)$

$= \lim\limits_{k \to \infty} \left(1 - \dfrac{1}{k+1} \right)$

$= 1$ converges

e.g.

$\sum_{n=1}^{\infty} \dfrac{1}{\sqrt{n}} = 1 + \dfrac{1}{\sqrt{2}} + \dfrac{1}{\sqrt{3}} + \cdots + \dfrac{1}{\sqrt{n}} + \cdots$ diverges, since $p = \dfrac{1}{2}$

4. Integral Test

If $a_n = f(n)$, and its corresponding function $f(x)$ is positive, continuous, and decreasing for $x \geq 1$, then

$$\sum_{n=1}^{\infty} a_n \quad \text{and} \quad \int_1^{\infty} f(x)dx$$

either both converge or both diverge.

Note: In general $\sum_{n=1}^{\infty} a_n \neq \int_1^{\infty} f(x)dx$, although if they both converge.

e.g. Determine the convergency of the series $\sum_{n=1}^{\infty} \dfrac{2n}{n^2+1}$.

$f(x) = \dfrac{2x}{x^2+1}$ is positive, continuous, and decreasing for $x \geq 1$.

$$\int_1^{\infty} \frac{2x}{x^2+1}dx = \lim_{b \to \infty} \int_1^b \frac{2x}{x^2+1}dx$$

$$= \lim_{b \to \infty} \int_1^b \frac{d(x^2+1)}{x^2+1}$$

$$= \lim_{b \to \infty} \ln(x^2+1) \Big|_1^b$$

$$= \infty$$

Therefore $\sum_{n=1}^{\infty} \dfrac{2n}{n^2+1}$ diverges.

e.g. Determine the convergency of the series $\sum_{n=1}^{\infty} \dfrac{\ln n}{n}$.

$f(x) = \dfrac{\ln x}{x}$ is positive, continuous, and decreasing for $x > e$.

(since $f' = \dfrac{1-\ln x}{x^2} < 0$ when $x > e$)

$$\int_1^{\infty} \frac{\ln x}{x}dx = \lim_{b \to \infty} \int_1^b \frac{\ln x}{x}dx$$

$$= \lim_{b \to \infty} \frac{(\ln x)^2}{2}\Big|_1^b$$

$$= \infty$$

Therefore $\sum_{n=1}^{\infty} \dfrac{\ln n}{n}$ is also divergent.

5. Comparison Test

Suppose that a_n and b_n are positive terms, and

if $\sum_{n=1}^{\infty} a_n$ converges and $b_n \leq a_n$ for all n, then $\sum_{n=1}^{\infty} b_n$ also converges.

if $\sum_{n=1}^{\infty} a_n$ diverges and $b_n \geq a_n$ for all n, then $\sum_{n=1}^{\infty} b_n$ also diverges.

e.g.   Determine the convergency of the series $\sum_{n=1}^{\infty} \dfrac{1}{2^n+1}$.

Since $\sum_{n=1}^{\infty} \dfrac{1}{2^n}$ converges   (geometric series, $r = \dfrac{1}{2} < 1$)

and $\dfrac{1}{2^n+1} < \dfrac{1}{2^n}$ for all n,

therefore $\sum_{n=1}^{\infty} \dfrac{1}{2^n+1}$ also converges.

e.g. $\sum_{n=1}^{\infty} \dfrac{\ln n}{n^3}$ converges since $\dfrac{\ln n}{n^3} < \dfrac{1}{n^2}$.

To prove convergence, often compare the series with $\dfrac{1}{n^2}$.

To prove divergence, often compare the series with $\dfrac{1}{n}$.

The Limit Comparison Test

If a_n and b_n are both positive and $\lim_{n \to \infty} \dfrac{a_n}{b_n} = L$, L is a finite number and not equal to zero,

then $\sum_{n=1}^{\infty} a_n$ and $\sum_{n=1}^{\infty} b_n$ either both converge or both diverge.

Note: a_n and b_n are infinitesimals of the same order, they have the same behavior.

e.g.   Determine the convergency of the series $\sum_{n=1}^{\infty} \dfrac{1}{3n^2+2n+1}$.

Since $\sum_{n=1}^{\infty} \dfrac{1}{n^2}$ converges   (P - series, $p = 2 > 1$, converges)

$$\lim_{n \to \infty} \frac{\frac{1}{3n^2+2n+1}}{\frac{1}{n^2}} = \frac{n^2}{3n^2+2n+1} = \frac{1}{3}$$

therefore $\sum_{n=1}^{\infty} \dfrac{1}{3n^2+2n+1}$ also converges.

e.g. $\sum_{n=2}^{\infty} \dfrac{1}{n\sqrt{n^2-1}}$ converges,

since $\lim\limits_{n\to\infty} \dfrac{\frac{1}{n\sqrt{n^2-1}}}{\frac{1}{n^2}} = \lim\limits_{n\to\infty} \dfrac{n}{\sqrt{n^2-1}} = \lim\limits_{n\to\infty} \sqrt{\dfrac{n^2}{n^2-1}} = 1$

Note: Geometric series $(r = \frac{1}{2})$ and P – series $(p = 1, p = 2)$ are usually used for comparison.

6. Alternating Series Test

An alternating series $\sum_{n=1}^{\infty}(-1)^{n-1}a_n = a_1 - a_2 + a_3 - a_4 \cdots$ $(a_n > 0)$

converges, if $a_n > a_{n+1}$ and $\lim\limits_{n\to\infty} a_n = 0.$

e.g.   Determine the convergency of the alternating harmonic series

$$\sum_{n=1}^{\infty}(-1)^{n-1}\frac{1}{n} = 1 - \frac{1}{2} + \frac{1}{3} - \frac{1}{4} \cdots$$

Since $\dfrac{1}{n} \geq \dfrac{1}{n+1}$ for all $n \geq 1$, and $\lim\limits_{n\to\infty} \dfrac{1}{n} = 0,$

the series is convergent.

e.g. $\sum_{n=1}^{\infty}(-1)^{n+1}\dfrac{\ln n}{n}$ converges, since it is alternating and decreasing.

When a partial sum S_n is used to estimate the convergent alternating series, the error is smaller than the first truncated term $|a_{n+1}|$.

e.g.   The error bound of S_5 for the convergent alternating series $\sum_{n=1}^{\infty}(-1)^{n+1}\dfrac{1}{\sqrt{n}}$ is $\dfrac{1}{\sqrt{6}}$.

7. The Absolute Convergence Test

If $\sum_{n=1}^{\infty} |a_n|$ converges, then $\sum_{n=1}^{\infty} a_n$ converges.

A series $\sum_{n=1}^{\infty} a_n$ is called conditionally convergent if it is convergent but not absolutely convergent.

8. Ratio Test – Very Important

$$\lim\limits_{n\to\infty} \left|\dfrac{a_{n+1}}{a_n}\right| = p$$

the series $\sum_{n=1}^{\infty} a_n$ is absolutely convergent if $p < 1$,
the series is divergent if $p > 1$
the test is inconclusive if $p = 1$

e.g.   Determine the convergency of the series

$$\sum_{n=1}^{\infty} \frac{e^n}{n!}$$

$$\lim_{n\to\infty} \left|\frac{a_{n+1}}{a_n}\right| = \lim_{n\to\infty} \frac{\frac{e^{n+1}}{(n+1)!}}{\frac{e^n}{n!}} = \lim_{n\to\infty} \frac{e^n \cdot e}{(n+1)n!} \cdot \frac{n!}{e^n}$$

$$= \lim_{n\to\infty} \frac{e}{n+1} = 0$$

Therefore the series converges.

e.g.   Determine the convergency of the series

$$\sum_{n=1}^{\infty} \frac{n^2}{2^n}$$

$$\lim_{n\to\infty} \left|\frac{a_{n+1}}{a_n}\right| = \lim_{n\to\infty} \frac{(n+1)^2}{2^{(n+1)}} \cdot \frac{2^n}{n^2} = \frac{1}{2}\lim_{n\to\infty}\left(\frac{n+1}{n}\right)^2 = \frac{1}{2}$$

Therefore the series converges.

9. The *n*th-Root Test

If $\lim_{n\to\infty} \sqrt[n]{a_n} = \rho$ exists for the series $\sum_{n=1}^{\infty} a_n$, then

(a) the series converges if $\rho < 1$;

(b) the series diverges if $\rho > 1$;

(c) the test is inconclusive if $\rho = 1$.

e.g.   For the previous example $\sum_{n=1}^{\infty} \frac{n^2}{2^n}$

$$\lim_{n\to\infty} \sqrt[n]{\frac{n^2}{2^n}} = \lim_{n\to\infty} \frac{(\sqrt[n]{n})^2}{2} \qquad (\text{since } \lim_{n\to\infty} \sqrt[n]{n} = 1)$$

$$= \frac{1}{2} < 1$$

Therefore the series converges.

Part II. Series with Variable Terms

1. Power Series

$$\sum_{n=0}^{\infty} c_n x^n = c_0 + c_1 x + c_2 x^2 + c_3 x^3 + \cdots + c_n x^n + \cdots$$

is called a power series in x centered at 0.

where x is the variable, coefficients c_n are different constants.

The ratio test is used to determine the interval of absolute convergence.
Here we denote the nth term $c_n x^n$ by u_n for convenience.

Let $\lim\limits_{n \to \infty} \left| \dfrac{u_{n+1}}{u_n} \right| = 1$, solve for $|x| = r$.

r is called the radius of convergence.
The series converges when $|x| < r$. $(-r < x < r)$
The series diverges when $|x| > r$. $(x < -r$ or $x > r)$
Endpoints x = -r and x = r must be tested separately.

Special case: If $\lim\limits_{n \to \infty} \left| \dfrac{u_{n+1}}{u_n} \right| = 0$, the series converges for all x.

If $\lim\limits_{n \to \infty} \left| \dfrac{u_{n+1}}{u_n} \right| = \infty$, the series diverges for all x except x = 0.

General Form of the Power Series

$$\sum_{n=0}^{\infty} c_n (x - a)^n = c_0 + c_1 (x - a) + c_2 (x - a)^2 + c_3 (x - a)^3 + \cdots + c_n (x - a)^n + \cdots$$

is called a power series in x centered at a .

Here we denote the nth term $c_n (x - a)^n$ by u_n for convenience.

Let $\lim\limits_{n \to \infty} \left| \dfrac{u_{n+1}}{u_n} \right| = 1$, solve for $|x - a| = r$.

The series converges when $|x - a| < r$. $(a - r < x < a + r)$
The series diverges when $|x - a| > r$. $(x < a - r$ or $x > a + r)$
Endpoints $x = a - r$ and $x = a + r$ must be tested separately.

A power series can be considered as a function in the variable x

$$f(x) = \sum_{n=0}^{\infty} c_n (x - a)^n$$

The domain of $f(x)$ is the convergent interval of the power series.

e.g. $\sum_{n=0}^{\infty} \dfrac{x^n}{n!} = 1 + x + \dfrac{x^2}{2!} + \dfrac{x^3}{3!} + \cdots$

$$\lim_{n \to \infty} \left| \dfrac{u_{n+1}}{u_n} \right| = \lim_{n \to \infty} \left| \dfrac{\frac{x^{n+1}}{(n+1)!}}{\frac{x^n}{n!}} \right| = \lim_{n \to \infty} \left| \dfrac{x}{n+1} \right| = 0$$

This series converges for all x.

e.g. $\sum_{n=1}^{\infty} \dfrac{x^n}{n} = x + \dfrac{x^2}{2} + \dfrac{x^3}{3} + \cdots$

$$\lim_{n \to \infty} \left| \dfrac{u_{n+1}}{u_n} \right| = \lim_{n \to \infty} \left| \dfrac{\frac{x^{n+1}}{(n+1)}}{\frac{x^n}{n}} \right| = \lim_{n \to \infty} \left| \dfrac{n\,x}{n+1} \right| = |x| \lim_{n \to \infty} \left| \dfrac{n}{n+1} \right| = |x|$$

$|x| < 1$, $-1 < x < 1$ converges.

Test the endpoints:

For $x = -1$,

$$\sum_{n=1}^{\infty} \dfrac{(-1)^n}{n} = -1 + \dfrac{1}{2} - \dfrac{1}{3} + \dfrac{1}{4} \cdots$$

which converges by the alternating series test.

For $x = 1$,

$$\sum_{n=1}^{\infty} \dfrac{1^n}{n} = 1 + \dfrac{1}{2} + \dfrac{1}{3} + \dfrac{1}{4} \cdots$$

which diverges (harmonic series).

Therefore the interval of convergence is $[-1, 1)$.

e.g. $\sum_{n=0}^{\infty} \dfrac{\sqrt{n}\,x^n}{2^n}$

$$\text{Let } \lim_{n \to \infty} \left| \dfrac{u_{n+1}}{u_n} \right| = \lim_{n \to \infty} \left| \dfrac{\sqrt{n+1}\,x^{n+1}}{2^{n+1}} \cdot \dfrac{2^n}{\sqrt{n}\,x^n} \right|$$

$$= \lim_{n \to \infty} \left| \dfrac{\sqrt{n+1}\,x}{2} \cdot \dfrac{1}{\sqrt{n}} \right| = \dfrac{|x|}{2} \lim_{n \to \infty} \left| \sqrt{\dfrac{n+1}{n}} \right|$$

$$= \dfrac{|x|}{2} = 1$$

Therefore $|x| = 2$

It converges when $|x| < 2$; it diverges when $|x| > 2$.

Test the endpoints $x = -2$ and $x = 2$.

$$\sum_{n=0}^{\infty} \frac{\sqrt{n}\,(-2)^n}{2^n} = \sum_{n=0}^{\infty}(-1)^n \sqrt{n} \text{ diverges.}$$

$$\sum_{n=0}^{\infty} \frac{\sqrt{n}\,(2)^n}{2^n} = \sum_{n=0}^{\infty} \sqrt{n} \text{ diverges.}$$

(Hint: $\lim_{n \to \infty} u_n \neq 0$)

Therefore the interval of convergence is $-2 < x < 2$.

e.g. $\qquad \sum_{n=0}^{\infty} \frac{\sqrt{n}\,(x-1)^n}{2^n}$

Use the result from the example above.

$-2 < x - 1 < 2$

$1 - 2 < x < 1 + 2$

$-1 < x < 3$

Term-by-Term Differentiation and Integration of the Power Series

A power series, its term-by-term differentiation and its term-by-term integration have the same radius of convergence.

Note: Their convergent intervals are not necessarily the same.

e.g. A geometric series is a power series with $c_n = 1$.

e.g. The geometric series

$$\sum_{n=0}^{\infty} x^n = 1 + x + x^2 + x^3 + \cdots + x^n + \cdots$$

Is a power series centered at 0.

Use the ratio test: $\lim_{n \to \infty} \left| \frac{u_{n+1}}{u_n} \right| = \lim_{n \to \infty} \left| \frac{x^{n+1}}{x^n} \right| = |x| = 1$

Therefore the radius of convergence is $r = 1$.

It converges when $|x| < 1$.

Test the endpoints $x = -1$ and $x = 1$. Both are divergent.

The interval of convergence is $-1 < x < 1$.

This is a geometric series with $a = 1, \quad r = x$

$$\lim_{n \to \infty} \Sigma_{n=0}^{\infty} x^n = \frac{a}{1-r} = \frac{1}{1-x}$$

$$1 + x + x^2 + x^3 + \cdots = \frac{1}{1-x}$$

Its term-by-term differentiation:

$$\lim_{n \to \infty} \Sigma_{n=1}^{\infty} nx^{n-1} = 1 + 2x + 3x^2 + 4x^3 + \cdots = \frac{1}{(1-x)^2}$$

It has the same radius of convergence $r = 1$.

Test the endpoints $x = -1$ and $x = 1$. Both are divergent.

e.g.   Since we know $\dfrac{1}{1-x} = 1 + x + x^2 + x^3 + \cdots$, substitute x by (-x)

$$\frac{1}{1+x} = 1 - x + x^2 - x^3 + \cdots$$

Its term-by-term integration:

$$\int \frac{1}{1+x} dx = \int (1 - x + x^2 - x^3 + \cdots) dx$$

$$\ln(1+x) + C = x - \frac{x^2}{2} + \frac{x^3}{3} - \frac{x^4}{4} \cdots$$

The series is zero when x = 0, therefore C = 0. Hence,

$$\ln(1+x) = x - \frac{x^2}{2} + \frac{x^3}{3} - \frac{x^4}{4} \cdots$$

Test the end points. $-1 < x \le 1$

2. Functions Represented as a Power Series

Taylor Series:

Let $f(x)$ be a function with derivatives of all orders on some interval containing a as an interior point.

Then $f(x)$ can be expressed by the **Taylor Series**:

$$f(x) = \sum_{n=0}^{\infty} \frac{f^{(n)}(a)}{n!}(x-a)^n = f(a) + f'(a)(x-a) + \frac{f''(a)}{2!}(x-a)^2 + \frac{f'''(a)}{3!}(x-a)^3 \cdots$$

The nth partial sum of the Taylor Series:

$$P_n(x) = \sum_{i=0}^{n} \frac{f^{(i)}(a)}{i!}(x-a)^i = f(a) + f'(a)(x-a) + \frac{f''(a)}{2!}(x-a)^2 \cdots + \frac{f^{(n)}(a)}{n!}(x-a)^n$$

is called the nth degree Taylor polynomial of f at a .

The first degree Taylor polynomial is the linear approximation. The higher-order Taylor polynomials provide better approximations.

When $a = 0$, the series is called **Maclaurin Series**:

$$f(x) = \sum_{n=0}^{\infty} \frac{f^{(n)}(0)}{n!}x^n = f(0) + f'(0)x + \frac{f''(0)}{2!}x^2 + \frac{f'''(0)}{3!}x^3 \cdots$$

e.g.   Find the Maclaurin Series for e^x

$$f(x) = e^x, \; f'(x) = e^x, f''(x) = e^x, \; \cdots, \; f^{(n)}(x) = e^x$$

$$f(0) = e^0 = 1, \; f'(0) = e^0 = 1, \; f''(0) = e^0 = 1, \; \cdots, \; f^{(n)}(0) = e^0 = 1$$

$$e^x = f(0) + f'(0)x + \frac{f''(0)}{2!}x^2 + \frac{f'''(0)}{3!}x^3 \cdots$$

$$= 1 + x + \frac{x^2}{2!} + \frac{x^3}{3!} + \cdots = \sum_{n=0}^{\infty} \frac{x^n}{n!}$$

e.g.   The third degree Taylor polynomial of e^x at 0 is

$$P_3(x) = 1 + x + \frac{x^2}{2!} + \frac{x^3}{3!}$$

e.g.   Find the Maclaurin series for $f(x) = e^x - x$ and show that $f(x)$ has a local minimum at $= 0$.

Since we know $e^x = 1 + x + \frac{x^2}{2!} + \frac{x^3}{3!} + \cdots$

$$e^x - x = 1 + \frac{x^2}{2!} + \frac{x^3}{3!} + \cdots \text{ , compare the coefficients}$$

$f'(0) = 0$ and $f''(0) = 1 > 0$, therefore $f(0)$ is a local minimum.

Lagrange Error Bound

If $f(x)$ is expanded as a Taylor Series about a and x is a number in the interval of convergence, then there is a number c between a and x such that the remainder $R_n(x)$ after the partial sum S_n is given by

$$R_n(x) = \frac{f^{(n+1)}(c)}{(n+1)!}(x-a)^{n+1}$$

$R_n(x)$ is called the Lagrange remainder.

In practice we use an upper bound M to replace $f^{(n+1)}(c)$ in error estimation.

$$|R_n(x)| \le \frac{M}{(n+1)!}|x-a|^{n+1}$$

If $f^{(n+1)}(x)$ is in the forms of $\sin x$ and $\cos x$, then we can use $|\sin x| = 1$, $|\cos x| = 1$ to compute M.

e.g.   The Maclaurin series for $\sin x = x - \frac{x^3}{3!} + \frac{x^5}{5!} - \frac{x^7}{7!} \cdots$. For what values of x can we

estimate $\sin x$ by $x - \frac{x^3}{3!}$ with an error no greater than 10^{-4} ?

$$|R_4| \le \frac{1}{5!}|x|^5$$

Solve $\frac{1}{5!}|x|^5 < 10^{-4}$, $|x| < \sqrt[5]{120 \times 10^{-4}} \approx 0.412$ rounded down to be safe.

e.g.   Express e by Taylor Series with an error of less than 10^{-3}.

$$e^x = \sum_{n=0}^{\infty}\frac{x^n}{n!} = 1 + x + \frac{x^2}{2!} + \frac{x^3}{3!} + \cdots + \frac{x^n}{n!} + R_n(x)$$

$$e = \sum_{n=0}^{\infty}\frac{1^n}{n!} = 1 + 1 + \frac{1}{2!} + \frac{1}{3!} + \cdots R_n(1) \qquad (a = 0, x = 1)$$

where $R_n(1) = \frac{f^{(n+1)}(c)}{(n+1)!} \cdot 1^n = \frac{e^c}{(n+1)!}$ $\qquad\qquad (f^{(n+1)}(x) = e^x)$

$e^0 < e^c < e^1$ for any c between 0 and 1

$1 < e^c < 3$ since we know $e < 3$

$\frac{1}{(n+1)!} < R_n(1) < \frac{3}{(n+1)!}$ $\qquad\qquad (M = 3, |x - a|^{n+1} = |1 - 0|^{n+1} = 1)$

Solve $\frac{3}{(n+1)!} < 10^{-3}$ $\qquad\qquad (n + 1 = 7, n = 6)$

(Hint: use calculator to do $3 \div 1 \div 2 \div 3 \div 4 \div \cdots$ until you see E^{-4}, which means 10^{-4}.)

$$e = \sum_{n=0}^{\infty}\frac{1^n}{n!} = 1 + 1 + \frac{1}{2!} + \frac{1}{3!} + \frac{1}{4!} + \frac{1}{5!} + \frac{1}{6!}$$

Alternating Series Estimation Theorem

For a convergent alternating series $\sum_{n=1}^{\infty}(-1)^{n-1}u_n$ $\qquad$ ($u_n > 0$)

$$|R_n(x)| \leq |u_{n+1}|$$

applied to alternating Taylor Series

$$|R_n(x)| \leq \left|\frac{f^{(n+1)}(a)}{(n+1)!}(x-a)^{n+1}\right|$$

e.g. $\ln(1+x) = x - \dfrac{x^2}{2} + \dfrac{x^3}{3} - \dfrac{x^4}{4} \cdots$

Determine the error in estimating $\ln(1.2)$ by the third degree Taylor polynomial.

$$|R_3(x)| \leq \left|\frac{x^4}{4}\right| = \left|\frac{0.2^4}{4}\right| = 4 \times 10^{-4} \qquad (x = 0.2)$$

Some Common Functions Defined by Power Series: $\qquad$ Interval of Convergence

$$\sin x = \sum_{n=0}^{\infty}(-1)^n \frac{x^{2n+1}}{(2n+1)!} = x - \frac{x^3}{3!} + \frac{x^5}{5!} - \frac{x^7}{7!} \cdots \qquad -\infty < x < \infty$$

$$\cos x = \sum_{n=0}^{\infty}(-1)^n \frac{x^{2n}}{(2n)!} = 1 - \frac{x^2}{2!} + \frac{x^4}{4!} - \frac{x^6}{6!} \cdots \qquad -\infty < x < \infty$$

$$\tan^{-1}x = \sum_{n=0}^{\infty}(-1)^n \frac{x^{2n+1}}{2n+1} = x - \frac{x^3}{3} + \frac{x^5}{5} - \frac{x^7}{7} \cdots \qquad -1 \leq x \leq 1$$

$$e^x = \sum_{n=0}^{\infty}\frac{x^n}{n!} = 1 + x + \frac{x^2}{2!} + \frac{x^3}{3!} + \cdots \qquad -\infty < x < \infty$$

$$\ln(1+x) = \sum_{n=1}^{\infty}(-1)^{n-1}\frac{x^n}{n} = x - \frac{x^2}{2} + \frac{x^3}{3} - \frac{x^4}{4} \cdots \qquad -1 < x \leq 1$$

$$\frac{1}{1-x} = \sum_{n=0}^{\infty}x^n = 1 + x + x^2 + x^3 + \cdots \qquad -1 < x < 1$$

$$\frac{1}{1+x} = \sum_{n=0}^{\infty}(-1)^n x^n = 1 - x + x^2 - x^3 + \cdots \qquad -1 < x < 1$$

e.g. Find the sum of $\sum_{n=0}^{\infty}(x+1)^n = 1 + (x+1) + (x+1)^2 + (x+1)^3 + \cdots$

Since we know $\dfrac{1}{1-x} = 1 + x + x^2 + x^3 + \cdots \qquad -1 < x < 1$

Use substitution: $-\dfrac{1}{x} = 1 + (x+1) + (x+1)^2 + (x+1)^3 + \cdots$

Interval of convergence: $-1 < x+1 < 1$, that is $-2 < x < 0$

1. Algebraic Formulas

$a^2 - b^2 = (a - b)(a + b)$
$a^3 - b^3 = (a - b)(a^2 + ab + b^2)$
$a^3 + b^3 = (a + b)(a^2 - ab + b^2)$
$a^n - b^n = (a - b)(a^{n-1} + a^{n-2}b + a^{n-3}b^2 \ldots\ldots + ab^{n-2} + b^{n-1})$

Binomial Expansions:
$(a + b)^2 = a^2 + 2ab + b^2$
$(a - b)^2 = a^2 - 2ab + b^2$
$(a + b)^3 = a^3 + 3a^2b + 3ab^2 + b^3$
$(a - b)^3 = a^3 - 3a^2b + 3ab^2 - b^3$
$(x + y)^n = {_n}C_0 x^n y^0 + {_n}C_1 x^{n-1} y^1 + {_n}C_2 x^{n-2} y^2 \ldots\ldots {_n}C_{n-1} x^1 y^{n-1} + {_n}C_n x^0 y^n$

Combination: $\quad {_n}C_r = \dfrac{{_n}P_r}{r!} \qquad (r \le n)$

Permutation: $\quad {_n}P_r = n(n - 1)(n - 2)(n - 3) \bullet\bullet\bullet\bullet\bullet\bullet (n - r + 1) \qquad$ (r factors)

Factorial: $\quad r! = r(r - 1)(r - 2)(r - 3) \bullet\bullet\bullet\bullet\bullet\bullet 3 \bullet 2 \bullet 1$

Completing Square: $x^2 + bx + (\dfrac{b}{2})^2 = (x + \dfrac{b}{2})^2$

Quadratic Formula: $ax^2 + bx + c = 0 \qquad\qquad$ where $a \ne 0$

$$x = \dfrac{-b \pm \sqrt{b^2 - 4ac}}{2a}$$

Laws of Exponents:

$a^m a^n = a^{m+n}, \quad (a^m)^n = a^{mn}, \quad a^{\frac{m}{n}} = \sqrt[n]{a^m}, \quad (ab)^n = a^n b^n$

$\dfrac{a^m}{a^n} = a^{m-n}, \quad a^{-n} = \dfrac{1}{a^n}, \quad a^0 = 1$

Laws of Logarithms:

$\quad \log_a AB = \log_a A + \log_a B, \quad \log_a A^n = n \bullet \log_a A, \quad \log_a \sqrt[n]{A} = \dfrac{1}{n} \log_a A$

$\quad \log_a \dfrac{A}{B} = \log_a A - \log_a B, \quad \log_a 1 = 0, \quad \log_a a = 1$

Change of Base Formula:

$\quad \log_a A = \dfrac{\log A}{\log a}, \qquad \log_a A = \dfrac{\ln A}{\ln a}, \quad a^x = e^{x \ln a}$

Absolute Values: $|x| = \sqrt{x^2}, \quad |ab| = |a||b|, \quad |a + b| \le |a| + |b|$

2. Geometry Formulas

Rectangle: Area $A = l \cdot w$ l: length   w: width
Parallelogram: Area $A = b \cdot h$ b: base   h: height
Triangle: Area $A = \dfrac{1}{2} b \cdot h$

Trapezoid: Area $A = \dfrac{b_1 + b_2}{2} \cdot h$

Circle: Area $A = \pi r^2$,   Circumference $C = 2\pi r = \pi d$,   r: radius   d: diameter

Sector of Circle: Area $A = \dfrac{1}{2} r^2 \theta$,   θ in radian

                      Arc Length $s = r\theta$

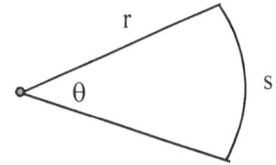

Right Prism: Volume: $V = B \cdot h$ B: Base Area , h: height
Pyramid: Volume: $V = \dfrac{1}{3} B \cdot h$

Right Circular Cylinder: Volume: $V = B \cdot h = \pi r^2 \cdot h$,
                          Lateral Area: $L = 2\pi r \cdot h$

Right Circular Cone: Volume: $V = \dfrac{1}{3} B \cdot h = \dfrac{1}{3} \pi r^2 \cdot h$,
                          Lateral Area: $L = \pi r \cdot l$ l: slant height

Sphere: Volume: $V = \dfrac{4}{3} \pi R^3$
           Surface Area: $SA = 4\pi R^2$

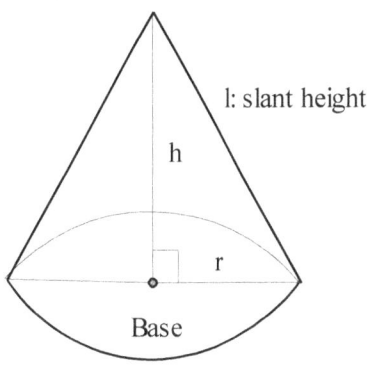

Frustum of Right Circular Cone:
       Lateral Area: $L = \pi (r_1 + r_2) \cdot l$

Frustum of Sphere:
       Lateral Arae: $L = 2\pi R h$

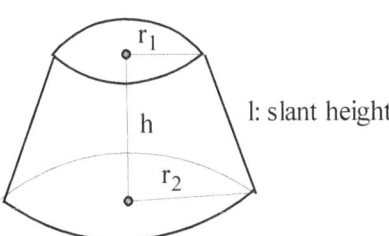

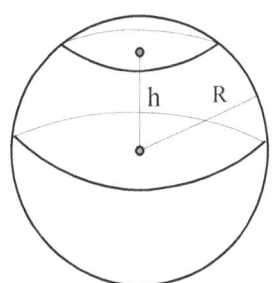

3. Trigonometry Formulas

Trigonometric Ratios and Basic Functions:

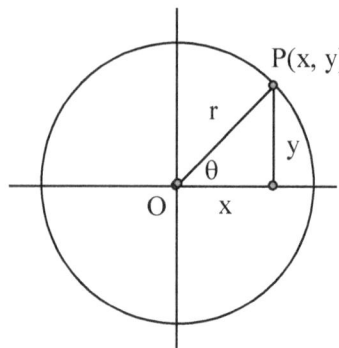

$$\sin \theta = \frac{y}{r}$$

$$\cos \theta = \frac{x}{r}$$

$$\tan \theta = \frac{y}{x}$$

$$r = \sqrt{x^2 + y^2}$$

Reciprocal Functions:

$$\cot A = \frac{1}{\tan A} \quad , \quad \sec A = \frac{1}{\cos A} \quad , \quad \csc A = \frac{1}{\sin A}$$

Exact Values to Remember:

θ (degree)	0°	30°	45°	60°	90°
θ (radian)	0	$\pi/6$	$\pi/4$	$\pi/3$	$\pi/2$
$\sin\theta$	0	$1/2$	$\sqrt{2}/2$	$\sqrt{3}/2$	1
$\cos\theta$	1	$\sqrt{3}/2$	$\sqrt{2}/2$	$1/2$	0
$\tan\theta$	0	$\sqrt{3}/3$	1	$\sqrt{3}$	undefined

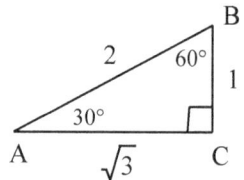
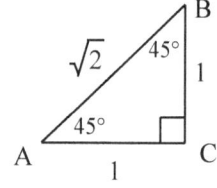

Pythagorean Triples:
   3, 4, 5 and 5, 12, 13

Trigonometric Identities:

$$\tan \theta = \frac{\sin \theta}{\cos \theta} \quad , \quad \cot \theta = \frac{\cos \theta}{\sin \theta} \quad ,$$

$$\sin^2 \theta + \cos^2 \theta = 1$$
$$\tan^2 \theta + 1 = \sec^2 \theta$$
$$\cot^2 \theta + 1 = \csc^2 \theta$$

Two Angles:
$$\sin(A + B) = \sin A \cos B + \cos A \sin B$$
$$\sin(A - B) = \sin A \cos B - \cos A \sin B$$
$$\cos(A + B) = \cos A \cos B - \sin A \sin B$$
$$\cos(A - B) = \cos A \cos B + \sin A \sin B$$

$$\tan(A + B) = \frac{\tan A + \tan B}{1 - \tan A \tan B}$$

$$\tan(A - B) = \frac{\tan A - \tan B}{1 + \tan A \tan B}$$

$$\sin A + \sin B = 2\sin\frac{A + B}{2}\cos\frac{A - B}{2}$$

$$\sin A - \sin B = 2\cos\frac{A + B}{2}\sin\frac{A - B}{2}$$

$$\cos A + \cos B = 2\cos\frac{A + B}{2}\cos\frac{A - B}{2}$$

$$\cos A - \cos B = -2\sin\frac{A + B}{2}\sin\frac{A - B}{2}$$

$$2\sin A \cos B = \sin(A + B) + \sin(A - B)$$

$$2\cos A \cos B = \cos(A + B) + \cos(A - B)$$

$$2\sin A \sin B = -\cos(A + B) + \cos(A - B)$$

Double Angles:

$$\cos2\theta = 2\cos^2\theta - 1 , \quad \cos2\theta = 1 - 2\sin^2\theta , \quad \cos2\theta = \cos^2\theta - \sin^2\theta$$

$$\cos^2\theta = \frac{1 + \cos2\theta}{2} , \quad \sin^2\theta = \frac{1 - \cos2\theta}{2}$$

$$\sin2\theta = 2\sin\theta\cos\theta , \quad \tan2\theta = \frac{2\tan\theta}{1 - \tan^2\theta}$$

Half Angles:

$$\sin\frac{\theta}{2} = \pm\sqrt{\frac{1 - \cos\theta}{2}} , \qquad 1 - \cos\theta = 2\sin^2\frac{\theta}{2}$$

$$\cos\frac{\theta}{2} = \pm\sqrt{\frac{1 + \cos\theta}{2}} , \qquad 1 + \cos\theta = 2\cos^2\frac{\theta}{2}$$

$$\tan\frac{\theta}{2} = \pm\sqrt{\frac{1 - \cos\theta}{1 + \cos\theta}} , \qquad \frac{1 - \cos\theta}{\sin\theta} = \tan\frac{\theta}{2}$$

The following identities are true for all values of θ:

$$\sin(-\theta) = -\sin\theta, \quad \cos(-\theta) = \cos\theta, \quad \tan(-\theta) = -\tan\theta$$
$$\sin(90° - \theta) = \cos\theta, \quad \cos(90° - \theta) = \sin\theta$$

Triangle Area $= \dfrac{1}{2}ab\sin C$

Law of Sines:

$$\frac{a}{\sin A} = \frac{b}{\sin B} = \frac{c}{\sin C} = 2R , \qquad \text{R is the radius of the circumscribed circle.}$$

Law of Cosines:

$$c^2 = a^2 + b^2 - 2ab\cos C$$
$$\cos C = \frac{a^2 + b^2 - c^2}{2ab}$$

Inverse Trigonometric Functions:

$y = \sin^{-1}x = \text{arc } \sin x$ \qquad implies \qquad $x = \sin y$, \qquad where $-\dfrac{\pi}{2} \leq y \leq \dfrac{\pi}{2}$

$y = \cos^{-1}x = \text{arc } \cos x$ \qquad implies \qquad $x = \cos y$, \qquad where $0 \leq y \leq \pi$

$y = \tan^{-1}x = \text{arc } \tan x$ \qquad implies \qquad $x = \tan y$, \qquad where $-\dfrac{\pi}{2} < y < \dfrac{\pi}{2}$

A2. Calculus Formulas

1. Rules of Differentiation

1. $\dfrac{d}{dx}(u \pm v) = \dfrac{du}{dx} \pm \dfrac{dv}{dx}$

2. $\dfrac{d}{dx}(uv) = v\dfrac{du}{dx} + u\dfrac{dv}{dx} = u'v + uv'$, $\quad \dfrac{d}{dx}(cu) = c\dfrac{du}{dx}$

3. $\dfrac{d}{dx}\left(\dfrac{u}{v}\right) = \dfrac{v\dfrac{du}{dx} - u\dfrac{dv}{dx}}{v^2} = \dfrac{u'v - uv'}{v^2}$

4. $\dfrac{dy}{dx} = \dfrac{dy}{du} \cdot \dfrac{du}{dx}$ $\qquad$ Chain Rule

5. $\dfrac{dy}{dx} = \dfrac{1}{\dfrac{dx}{dy}}$, $\quad \dfrac{dy}{dx} = \dfrac{\dfrac{dy}{dt}}{\dfrac{dx}{dt}}$, $\quad \dfrac{d^2y}{dx^2} = \dfrac{d\left(\dfrac{dy}{dx}\right)/dt}{dx/dt}$, $\quad \dfrac{dy}{dx} = y\dfrac{d}{dx}(\ln y)$

Rules of Integation

1. If $\int f(x)dx = F(x) + C$, then $\int f(u)dx = F(u) + C$ $\qquad$ where u = u(x) is differentiable
2. $\int c\,f(x)dx = c\int f(x)dx$
3. $\int [f(x) \pm g(x)]dx = \int f(x)dx \pm \int g(x)dx$
4. $\int u \cdot v'dx = \int udv = uv - \int vdu$ $\qquad$ Integration by Parts
   In general u: $\ln x$, x^n , $\sin^{-1}$, $\cos^{-1}$, $\tan^{-1}$; $\quad v' = \sin x$, $\cos x$, e^x
5. $\dfrac{d}{dx}\int f(x)dx = f(x)$ $\qquad$ or $\quad d\int f(x)dx = f(x)dx$
6. $\int \dfrac{d}{dx}f(x)dx = f(x) + C$ $\quad$ or $\quad \int df(x) = f(x) + C$

Rules of Definite Integration

1. $\int_a^b f(x)dx = F(b) - F(a)$ $\qquad$ where $F'(x) = f(x)$
2. $\int_a^a f(x)dx = 0$
3. $\int_a^b f(x)dx = -\int_b^a f(x)dx$
4. $\int_a^b f(x)dx = \int_a^c f(x)dx + \int_c^b f(x)dx$ $\qquad$ (a, b, c can be in any order)
5. $\int_a^b [f(x) \pm g(x)]dx = \int_a^b f(x)dx \pm \int_a^b g(x)dx$
6. $\int_a^b cf(x)dx = c\int_a^b f(x)dx$ $\qquad$ where c is any constant
7. $\int_a^b f(u) \cdot u'dx = \int_{u(a)}^{u(b)} f(u)du$ $\qquad$ where u = u(x) Substitution Formula
8. $\dfrac{d}{dx}\int_a^x f(t)\,dt = f(x)$, $\quad \dfrac{d}{dx}\int_a^{g(x)} f(t)\,dt = f(g(x)) \cdot \dfrac{d}{dx}g(x)$
9. $\int f(x)dx = \int_a^x f(t)dt + C = F(x) + C$ $\qquad$ where $F'(x) = f(x)$
10. $\int_a^\infty f(x)dx = \lim_{b\to\infty} \int_a^b f(x)dx$

2. Basic Derivative Table

a and c are constant; u(x) and v(x) are differentiable

1. $\dfrac{d}{dx} c = 0$

2. $\dfrac{d}{dx} x^n = nx^{n-1}$, $\dfrac{d}{dx}\left(\dfrac{1}{x}\right) = -\dfrac{1}{x^2}$, $\dfrac{d}{dx}\sqrt{x} = \dfrac{1}{2\sqrt{x}}$

3. $\dfrac{d}{dx} e^x = e^x$, $\dfrac{d}{dx} a^x = a^x \ln a$, $\dfrac{d}{dx} e^{-x} = -e^{-x}$

4. $\dfrac{d}{dx}\ln x = \dfrac{1}{x}$ $\dfrac{d}{dx}\log_a x = \dfrac{1}{x}\log_a e = \dfrac{1}{x \ln a}$,

5. $\dfrac{d}{dx}\sin x = \cos x$, $\dfrac{d}{dx}\sin^{-1}x = \dfrac{1}{\sqrt{1-x^2}}$

6. $\dfrac{d}{dx}\cos x = -\sin x$, $\dfrac{d}{dx}\cos^{-1}x = -\dfrac{1}{\sqrt{1-x^2}}$

7. $\dfrac{d}{dx}\tan x = \sec^2 x$, $\dfrac{d}{dx}\tan^{-1}x = \dfrac{1}{1+x^2}$

8. $\dfrac{d}{dx}\cot x = -\csc^2 x$, $\dfrac{d}{dx}\cot^{-1}x = -\dfrac{1}{1+x^2}$

9. $\dfrac{d}{dx}\sec x = \sec x \tan x$, $\dfrac{d}{dx}\sec^{-1}x = \dfrac{1}{|x|\sqrt{x^2-1}}$

10. $\dfrac{d}{dx}\csc x = -\csc x \cot x$, $\dfrac{d}{dx}\csc^{-1}x = -\dfrac{1}{|x|\sqrt{x^2-1}}$

3. Basic Integral Table

1. $\int u^n\, du = \dfrac{u^{n+1}}{n+1} + C \qquad (n \neq -1)\,,$ $\int \dfrac{du}{u} = \ln|u| + C$

   $\int u\, du = \dfrac{u^2}{2} + C\,,$ $\int \dfrac{du}{u^2} = -\dfrac{1}{u} + C\,,$ $\int \dfrac{du}{\sqrt{u}} = 2\sqrt{u} + C$

2. $\int e^u\, du = e^u + C\,,$ $\int a^u\, du = \dfrac{a^u}{\ln a} + C$

   $\int \ln u\, du = u\ln u - u + C$

3. $\int \sin u\, du = -\cos u + C$

   $\int \dfrac{du}{\sin u} = \ln|\csc u - \cot u| + C\,,$ $\int \dfrac{du}{\sin^2 u} = -\cot u + C$

5. $\int \cos u\, du = \sin u + C$

   $\int \dfrac{du}{\cos u} = \ln|\sec u + \tan u| + C\,,$ $\int \dfrac{du}{\cos^2 u} = \tan u + C$

7. $\int \tan u\, du = \ln|\sec u| + C = -\ln|\cos u| + C$

8. $\int \cot u\, du = \ln|\sin u| + C$

9. $\int \sec u\, du = \ln|\sec u + \tan u| + C$

   $\int \sec^2 u\, du = \tan u + C$

10. $\int \csc u\, du = \ln|\csc u - \cot u| + C$

   $\int \csc^2 u\, du = -\cot u + C$

11. $\int \sec u \tan u\, du = \sec u + C$

12. $\int \csc u \cot u\, du = -\csc u + C$

13. $\int \dfrac{du}{u^2+1} = \tan^{-1} u + C\,,$ $\int \dfrac{du}{u^2+a^2} = \dfrac{1}{a}\tan^{-1}\dfrac{u}{a} + C$

   $\int \dfrac{du}{u^2-a^2} = \dfrac{1}{2a}\ln\left|\dfrac{u-a}{u+a}\right| + C$

14. $\int \dfrac{du}{\sqrt{1-u^2}} = \sin^{-1} u + C\,,$ $\int \dfrac{du}{\sqrt{a^2-u^2}} = \sin^{-1}\dfrac{u}{a} + C\,,$

   $\int \dfrac{du}{\sqrt{u^2 \pm a^2}} = \ln\left|u + \sqrt{u^2 \pm a^2}\right| + C$

15. $\int \dfrac{du}{u\sqrt{u^2-1}} = \sec^{-1}|u| + C\,,$ $\int \dfrac{du}{u\sqrt{u^2-a^2}} = \dfrac{1}{a}\sec^{-1}\left|\dfrac{u}{a}\right| + C\,,$

4. Infinite Series

1. Geometric Series
$$\sum_{n=1}^{\infty} ar^{n-1} = a + ar + ar^2 + \cdots + ar^{n-1} + \cdots$$
   if $|r| < 1$, the series converges ;
   if $|r| \geq 1$, the series diverges.

$$S_n = \frac{a(1-r^n)}{(1-r)} \qquad r \neq 1 , \qquad\qquad \lim_{n\to\infty} S_n = \frac{a}{(1-r)} \quad |r| < 1$$

$$\sum_{n=0}^{\infty} x^n = 1 + x + x^2 + \cdots + x^n + \cdots = \frac{1}{1-x} \quad \text{converges for } -1 < x < 1$$

2. Ratio Test
   For all positive terms:
$$\lim_{x\to\infty} \frac{a_{n+1}}{a_n} = p$$
        if $p < 1$, the series converges
        if $p > 1$, the series diverges
        (if $p = 1$, the test is insufficient)

3. Alternating Series
$$\sum_{n=1}^{\infty} (-1)^{n+1} a_n$$
   Converges if $a_n > a_{n+1}$ and $\lim_{n\to\infty} a_n = 0$ (decreasing to zero)

4. Integral Test
   If f is positive, continuous, and decreasing, then
$$\sum_{n=1}^{\infty} f(n) \text{ and } \int_1^{\infty} f(x)dx$$
   either both converge or both diverge.

5. P – Series
$$\sum_{n=1}^{\infty} \frac{1}{n^p} = 1 + \frac{1}{2^p} + \frac{1}{3^p} + \cdots + \frac{1}{n^p} + \cdots \qquad\qquad \begin{array}{l}\text{if } p > 1 , \text{ converges} \\ \text{if } p \leq 1 , \text{ diverges}\end{array}$$

   Harmonic Series: $\sum_{n=1}^{\infty} \frac{1}{n} = 1 + \frac{1}{2} + \frac{1}{3} + \cdots + \frac{1}{n} + \cdots$ diverges

6. Power Series
$$\sum_{n=0}^{\infty} a_n x^n = a_0 + a_1 x + a_2 x^2 + \cdots + a_n x^n + \cdots$$

   Use Ratio Test: $\lim_{n\to\infty} \left| \frac{a_{n+1} x^{n+1}}{a_n x^n} \right| = 1$, solve it for the radius of convergence r.
        $|x| < r$ or $-r < x < r$, converges ;
        $|x| > r$ or $x < -r$ or $x > r$, diverges .
        (endpoints $x = -r$ and $x = r$ must be tested separately)

Some Common Functions Defined by Power Series: Interval of Convergence

$$\sin x = x - \frac{x^3}{3!} + \frac{x^5}{5!} - \frac{x^7}{7!} \cdots \qquad\qquad -\infty < x < \infty$$

$$\cos x = 1 - \frac{x^2}{2!} + \frac{x^4}{4!} - \frac{x^6}{6!} \cdots \qquad\qquad -\infty < x < \infty$$

$$\tan^{-1} x = x - \frac{x^3}{3} + \frac{x^5}{5} - \frac{x^7}{7} \cdots \qquad\qquad -1 \leq x \leq 1$$

$$e^x = 1 + x + \frac{x^2}{2!} + \frac{x^3}{3!} + \cdots \qquad\qquad -\infty < x < \infty$$

$$\ln(1 + x) = x - \frac{x^2}{2} + \frac{x^3}{3} - \frac{x^4}{4} \cdots \qquad\qquad -1 < x \leq 1$$

Part 1. Basic Operations

1. Clear the Memory:
[2nd] [MEM] 7: Reset ... [ENTER]
1: All Ram ...[ENTER]   2: Reset [ENTER]
                Ram Cleared
[2nd] [MEM] 7: Reset ... [ENTER]
2: Defaults ...[ENTER]   2: Reset [ENTER]
                Defaults Set

Return to Home Screen:
[2nd] [QUIT]

2. Graph Functions

e.g.   Graph $y = -x^2 + 4$
 [Y =] [(-)] [X, T, θ, n] [x^2] [+] [4] [GRAPH]

e.g.   Graph $y = |x - 4|$
   [Y =] [MATH] NUM / 1: abs [ENTER]
       [X, T, θ, n] [-] [4] [ )] [GRAPH]

To graph the inverse function of f(x), [2nd] [DRAW] 8:
DrawInv [ENTER] f(x) [ENTER]

To graph parametric and polar equations reset [MODE]

Zoom Menu:
To have a better view of the graph:
 [ZOOM] 6: ZStandard (- 10 < x < 10 ; - 10 < y < 10)
 [ZOOM] 4: ZDecimal   (to see friendly windows)
 [ZOOM] 0: ZoomFit   (to see complete graphs)
 [ZOOM] 2: Zoom In   (to see details around cursor)

 [ZOOM] 7: ZTrig   (for Trigonometry ; $X_{scl} = \frac{\pi}{2}$)

 [ZOOM] 9: ZoomStat   (for Statistics)
e.g. Graph $y = \sin 2x$
        [Y =] [sin] [2] [X, T, θ, n] [ZOOM] [7]

Change Window Dimensions:
e.g.   Graph $y = -3x^2 + 12x + 5$
 [Y =] [(-)] [3] [X, T, θ, n] [x^2] [+] [1] [2]
 [X, T, θ, n] [+] [5] [ZOOM] [6]
   To see the complete graph:
        [WINDOW] Y_{max} = 20 [ENTER] [GRAPH]

Trace:
To see the y values vary with x values:
       [TRACE]
To find the value of y at a specific value of x:
       [2nd] [CALC] 1: value [ENTER]

* TI-84 Plus graphing calculator is used for the examples.

3. Table of a Function
e.g.   Display the table of $y = x^2$
       [Y =] [X, T, θ, n] [x^2]
       [2nd] [TABLE]
   To change the x increment:
       [2nd] [TBLSET] ΔTbl

4. Calculations

Solve Equations:

e.g.   Solve $x^2 - 9 = 0$
 (1) graph $y = x^2 - 9$
 (2) [2nd] [CALC] 2: zero [ENTER]
 (3) move the cursor to set the Left Bound [ENTER]
     and the Right Bound [ENTER] of the x - intercept,
     then Guess [ENTER]
         Zero x = - 3   y = 0
 (4) repeat (3) to find the other zero
         Zero x = 3   y = 0

Solve the System of Equations:

e.g. solve system   xy = 8
                 y = x + 2
 (1) rewrite the first Eq. as $y = \frac{8}{x}$
 (2) graph those two functions
 (3) [2nd] [CALC] 5: intersect [ENTER]
 (4) Y_1 = 8 /X
     First Curve?
     move the cursor to the intersection [ENTER]
     Y_2 = X + 2
     Second Curve ? [ENTER]
     [GUESS] [ENTER]
     Intersection X = 2   Y = 4
 (5) repeat (4) to find the other intersection:
         X = -4   Y = -2

Maximum and Minimum:

e.g.   Find the maximum or minimum of the function
         $y = x^2 - 6x + 3$
 (1) graph $y = x^2 - 6x + 3$
 (2) [2nd] [CALC] 3: minimum [ENTER]
 (3) move the cursor to set the Left Bound [ENTER]
     and the Right Bound [ENTER] of the minimum,
     then Guess [ENTER]
         Minimum x = 3   y = -6
     when x = 3 the function has a minimum of -6 .

Part 2. Calculus Applications

1. Graph Piecewise Defined Functions

e.g.

$$f(x) = \begin{cases} x^2 & \text{for } -3 < x \le 2 \\ \\ 8 - x & \text{for } x > 2 \end{cases}$$

Graph $Y_1 = x^2 / (-3 < x)(x \le 2)$
            $Y_2 = (8 - x) / (x > 2)$

[Y =] Y$_1$ = [X, T, θ, n] [x^2] [÷] [( ] [(-)] [3]
            [2nd] [TEST] TEST/ 5: < [ENTER]
            [X, T, θ, n] [ )] [( ] [X, T, θ, n]
            [2nd] [TEST] TEST/ 6: ≤ [ENTER]
            [2] [ )]
         Y$_2$ = [( ] [8] [-] [X, T, θ, n] [ )] [÷]
            [( ] [X, T, θ, n] [2nd] [TEST]
            TEST/ 3: > [ENTER] [2] [ )]

Display the graphs: [ZOOM] [6]

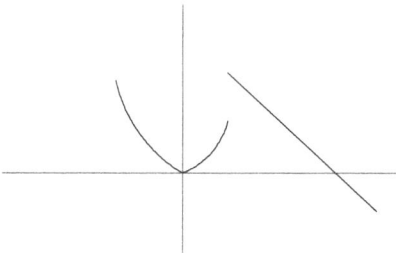

Note: Graphing calculator can not show the endpoints. We can not determine whether the endpoints are defined or not. Therefore the graphing calculator can only show the discontinuities that are unremovable.

2. Find the Numerical Derivatives

e.g. Find the derivative of $f(x) = x^2 + 2x - 1$ at $x = 3$.
    (1) graph $y = x^2 + 2x - 1$
    (2) [2nd] [CALC] 6: dy/dx [ENTER]
        [3] [ENTER]

    Display: dy/dx = 8

or [MATH] 8: nDeriv ($x^2 + 2x - 1$, x, 3) [ENTER]
(To see the derivative curve, replace 3 by x)

3. Find the Numerical Integrals

e.g. Find the displacement of a particle from t = 1
     to t = 8. Its velocity is v(t) = lnt - t + 4.

$$\int_1^8 v(t)\,dt = \int_1^8 (\ln t - t + 4)dt = 6.136$$

(1) graph $y = \ln x - x + 4$
(2) [2nd] [CALC] 7: S f(x)dx [ENTER]
    Lower Limit? [1] [ENTER]
    Upper Limit? [8] [ENTER]

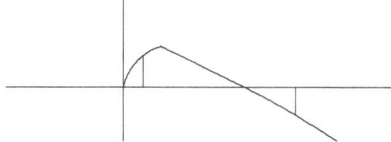

or [MATH] 9: fnInt (lnx - x + 4, x, 1, 8) [ENTER]
(To see the integral curve, replace 8 by x)

Find the disdance traveled by particle

$$\int_1^8 |v(t)|\,dt = \int_1^8 |(\ln t - t + 4)|dt = 10.418$$

absolute value: [MATH] NUM 1: abs ([ENTER]

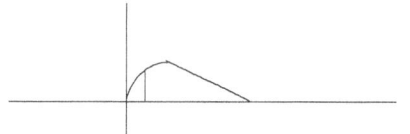

4. Sequence and Series:

e.g. $e = \sum \dfrac{1}{n!}$ n = 0, 1, 2, 3, 4, $\cdots$

      $= 1 + 1 + \dfrac{1}{2!} + \dfrac{1}{3!} + \dfrac{1}{4!} \cdots$

Find the first 10 terms and the sum.
Step 1: Enter the sequence and store it in L$_1$
[2nd] [LIST] OPS / 5: seq [ENTER] [1] [÷]
[X, T, θ, n] [MATH] PRB / 4: ! [ENTER] [,]
[X, T, θ, n] [,] [0] [,] [9] [ )]
[STO >] [2nd] [L$_1$] [ENTER]
 Display: seq(1 / x ! , x , 0 , 9) → L$_1$
{ 1   1   .5   .1666
To see the rest of the terms, press the arrow key [>]

Step 2: Find the sum --- the series.
[2nd] [LIST] MATH / 5: sum [ENTER] [2nd]
[L$_1$] [ )] [ENTER]
    Display: sum (L$_1$)   2.718281526

Also Available

Teacher's Choice
Math Regents Review   ISBN: 9781450562843

This book covers all the topics of high school math: Algebra, Geometry and Trigonometry. It is an ideal reference book for math teachers and college students.

Student's Choice
Regents Review Integrated Algebra   ISBN: 9781453880982

Student's Choice
Regents Review Geometry   ISBN: 9781453709993

Student's Choice
Regents Review Algebra 2/Trigonometry   ISBN: 9781460983874

These books are structured in three parts: Review, Practice and Complete Answers to Real Regents Questions

www.ingramcontent.com/pod-product-compliance
Lightning Source LLC
Chambersburg PA
CBHW081459170526
45166CB00008B/2480